GAIA VINCE

Transcendence

*How Humans Evolved through Fire,
Language, Beauty and Time*

ALLEN LANE
an imprint of
PENGUIN BOOKS

ALLEN LANE

UK | USA | Canada | Ireland | Australia
India | New Zealand | South Africa

Allen Lane is part of the Penguin Random House group of companies
whose addresses can be found at global.penguinrandomhouse.com

First published in the USA by Basic Books 2019
First published in Great Britain by Allen Lane 2019
002

Copyright © Gaia Vince, 2019

The moral right of the author has been asserted

Set in 10.5/14 pt Sabon LT Std
Typeset by Jouve (UK), Milton Keynes
Printed and bound in Great Britain by Clays Ltd, Elcograf S.p.A.

A CIP catalogue record for this book is available from the British Library

ISBN: 978–0–241–28111–6

*For my parents,
whose fault it all is
by nature or nurture*

Contents

Introduction

When Neil Harbisson went to renew his UK passport in 2004, there was a problem with the photograph he had provided. It should contain 'no other people or objects. No hats, no infant dummies, no tinted glasses'.

The regulations didn't say anything about antennae.

Nevertheless, he was told to remove the accessory from his head and resubmit his application. Harbisson explained that his antenna was not an accessory, but a part of him – 'an extension of his brain' – and anyway, he couldn't remove it as it had been surgically implanted. The passport was issued.

That is how Harbisson became the world's first officially recognized cyborg.

Harbisson describes himself as the first 'trans-species' person. Through a technological adaptation he has evolved into something else – something beyond a biological human, something beyond nature.

Harbisson now has extrasensory perception: he can hear colours through his antenna. He had been born a biologically compromised human, unable to see in colour, as the result of a rare genetic condition called achromatopsia. Through his eyes, the world appears in shades of grey. As a 21-year-old art student, he collaborated with a couple of software programmers and a musician to develop an electronic device that would allow him to sense colours as musical notes and chords. In 2004, after a difficult search, he found a doctor who, on condition of anonymity, would implant the device.

The antenna is a black flexible wand that emerges from somewhere under his straw-blond hair at the back of his head, and protrudes up and over his forehead. Harbisson wears his hair in a severe bowl-shaped cut, shaved up at the back, so that it resembles a helmet, further blurring the line between the biological and artificial. At the front of the antenna is an electronic 'eye' that detects the colours of the objects around him and transmits these light frequencies to a chip implanted in his skull. There, these impulses are converted to sound frequencies and Harbisson hears the colours of the world through the bones of his head.

Initially, he struggled to make sense of the overwhelming colour information flooding his mind, and to discern and distinguish colour sounds by their names. But, 15 years on, he lives in a fabulous Technicolor symphony – he even dreams in colour. His biological brain has merged so completely with the electronic software that he now experiences sounds, speech, bleeps and other noise as colour. He began painting people's voices and musical compositions, from Mozart to Lady Gaga. Then he decided to expand his palette beyond the human range. Now, Harbisson can perceive ultraviolet and infrared, so that he can 'see' objects in the dark, appreciate patterns invisible to the rest of us unenhanced humans, and can even spot the UV markers left on tree trunks that animals produce in their urine. He has also upgraded his chip to allow Internet access, so he can connect to satellites and receive colours from external devices. It's an organ that is still evolving, Harbisson says.

In 2018, he had compass components fixed inside his knees to allow him to sense the earth's magnetic field, and his next implant will be a crown-like device he has designed, which he describes as an organ for time. It will span his head, producing a heat spot that will revolve around his skull in 24-hour cycles, allowing him to perceive time – to, in effect, sense the earth's rotation. Once his brain has accepted and integrated the new organ, Harbisson hopes to be able to stretch or speed his perception of time by altering the speed of the heat spot's motion. If he wants a moment to last longer, for example, he will be able to slow the heat motion. In this way, he might even be able to change his sensation of ageing, manipulating his relative experience of time, to live to 170. 'In the same way that we can create optical illusions, because we have an organ for the sense of sight, I think we can create time illusions if we have an organ for the sense of time,' he explains.

The term 'cyborg' was coined in 1960* by American scientists Man-
fred Clynes and Nathan Kline, who were describing their vision of an
enhanced human that could survive in an extraterrestrial environment.
Now, this fiction has become reality for Harbisson, and also for the
hundreds of millions of people who rely on contact lenses, cochlear
implants, artificial heart valves, and a range of other bionic aids to
enhance their natural abilities. Whether integrated into our bodies or
not, our tools and gadgets give us exceptional powers: we can fly without wings, dive without gills, be resuscitated after death, escape our
planet to set foot on the moon. More prosaically, they are the blades that
improve on our teeth and nails' ability to cut our food and the soled
shoes that help our feet run fast on stony ground. In truth, we are all
cyborgs, for none of us can survive without our technological inventions.

But to think of us simply as a sort of smarter chimp with cool tools is
to miss what is truly extraordinary about us and the way we operate on
this planet. Yes, we have evolved incredibly diverse and complex gadgets, but so too have we evolved languages, artworks, societies, genes,
landscapes, foods, belief systems and so much more. Indeed, we have
evolved an entire human world – a societal operating system – without
which Harbisson's antennae would not only not exist, but also be point-
less. For it is our human world that gives our technologies meaning and
drives their invention. We are so much more than evolved cyborgs.

I presume you are not reading this while perched naked in a tree in
the Congo jungle. You are, like me, wearing clothes, processed from
plants grown thousands of miles away, woven, dyed, cut and stitched
by different hands, aided by several machines, to somebody's design
somewhere else, shipped to another place, priced and marketed by other
people, working to various orders, and eventually, several steps later, of
your own unique volition, wrapping your skin as wonderfully as fur.
Perhaps you are sitting on a plastic chair formed out of the processed carcasses of long-dead sea creatures, held up by steel legs generated from
mined rocks, blasted and refined and assembled in multiple steps by
teams of people independently fashioning a structure that was devised
and altered over millennia, millions of times.

* The idea is at least a century older, though: in 1843, the horror writer Edgar Allan
Poe described a man with extensive prostheses.

Wherever you are, you are generating in your mind these words that I have written as though I were speaking them into your ear. My mind is directly connecting with your mind now, even though I wrote this in another time and place, perhaps in another language. It's possible that I'm no longer even alive.

You are smart but, when alone, you are fairly powerless. We live our lives utterly dependent on countless strangers for our survival. Men and women have toiled to make and assemble the constituents of my lunch, clothes, furniture, house, road, city, state and world beyond me. These many cooperating, collaborating strangers have themselves relied on thousands upon thousands of other people, living and dead, to shape the lives they lead. And yet there is no contract, no plan and no common purpose to our 7 billion lives.

If it seems incredible that everything we see now – all the busyness and industry of billions of people living seemingly autonomous yet utterly interdependent lives – could have arisen without any plan, then consider this: our superb working body, from its eyes to its toenails to its consciously aware brain, emerged similarly from a single cell, in a matter of weeks. As a fertilized egg cell begins to grow and divide, the one cell becomes a mass of pluripotent cells, meaning they have the potential to be any type of cell in the body, depending on their biological developing bath. Thus, a cell that finds itself by chance on the outer part of the ball may end up developing into a nerve cell in the spinal cord; another cell, depending on its developing bath, will become a heart cell. Evolution has created a mechanism whereby a functioning system of cooperating organs and cells – a human being – can be built from a simple cell.

We are each of us individuals with our own motivations and desires, and yet much of our autonomy is an illusion. We are formed in a cultural 'developing bath' that we will ourselves then fashion and maintain – a grand social project without direction or goal that has nevertheless produced the most successful species on Earth.

Humans now live longer and better than ever before, and we are the most populous big animal on Earth. Meanwhile, our closest living relatives, the now endangered chimpanzees, continue to live as they have for millions of years. We are not like the other animals, yet we evolved through the same process. What are we then?

This question fascinated me and I set out to understand our

exceptional nature and what alchemy created humanity – this planet-altering force of nature – out of an ape.

What follows is a remarkable evolution story that has captivated me utterly. It all rests on a special relationship between the evolution of our genes, environment and culture, which I call our human evolutionary triad. This mutually reinforcing triad creates the extraordinary nature of us, a species with the ability to be not simply the objects of a transformative cosmos, but agents of our own transformation. We have diverged from the evolutionary path taken by all other animals, and, right now, we are on the cusp of becoming something grander and more marvellous. As the environment that created us is transformed by us we are beginning our greatest transcendence.

Let me explain.

We are Earthly beings – Earth-conceived and Earth-born. The role of our planetary home in making a species that would itself reshape that planet is little appreciated, and yet the environment made us the people we are today. After all, it is in response to our environment that we walk on two legs, speak tonal languages, have immunity to the flu virus and developed culture. So, my story begins with the geological origins of our *Genesis*. All life is formed of the stuff of the universe, and we humans are fundamentally a microcosm of the grand cosmos. The calcium in the limestone cliffs supporting our coastlines is also in the bones that support us internally – both owe their provenance to the stars. The water coursing through our planet's rivers, much like our internal rivers of blood, has its origin in comets.

Humans emerged, like every other life form, through the process of biological evolution. Species change over time because randomly occurring genetic differences accumulate within populations over generations. Organisms whose genes make them more successful in their environment are more likely to survive and reproduce, thus passing their genes to subsequent generations. In this way, biology adapts in response to environmental pressures, and species have gradually evolved to exploit every habitat on Earth.*

---

* Living beings also respond and adapt to environmental challenges during their lives with physiological and behavioural changes, most of which are genetically programmed, inherited and instinctive.

Our intelligent, social ancestors also evolved adaptations to survive in their environment, which, for our early hominid forebears, was a tropical forest habitat, and one of these adaptations was culture. 'Culture' has so many different interpretations, but when I use the term, I am referring to learned information expressed in our tools, technologies and behaviours. Human culture relies on our ability to learn from others and to express this knowledge ourselves. We are not the only species to have evolved culture, but ours is far more flexible: it is cumulative and it evolves. Human cumulative culture ratchets up in complexity and diversity over generations to generate ever more efficient solutions to life's challenges.

Cumulative cultural evolution has proven a game changer in the story of life on Earth. Instead of our evolution being driven solely by changes to environment and genes, culture also plays its part. Cultural evolution shares much with biological evolution. Genetic evolution relies on variation, transmission and differential survival. All three are there with cultural evolution. The main difference is that, in biological evolution, they are operational mostly at the level of the individual, whereas for culture, group selection is more important than individual selection, as we shall see. It is our collective human culture, even more than our individual intelligence, that makes us smart.

We weren't the only human species to go down this evolutionary route – and we will visit our cousins – but we are the only ones to have survived. Hundreds of thousands of years ago, we began to escape our original environmental cradle by using our culture to overcome the physical and biological limitations that trap other species into uncreative lives. Our extraordinary evolution is driven by four key agents, which I describe in the following sections: *Fire*, *Word*, *Beauty* and *Time*.

Fire describes how we outsource our energy costs to escape our biological limitations and extend our physical capabilities. *Word* investigates the role of information in our success: the use of language to accurately transmit and store complex cultural knowledge and communicate ideas between minds. Language is a social glue that binds us with joint stories, and enables us to make better predictions and decide who to trust based on their reputations. *Beauty* encapsulates the importance of meaning in our activities, which enables us to coalesce around shared beliefs and identities. Our artistic expression produces cultural speciation – tribalism between and within our societies – but also enables the trade in resources, genes and ideas that prevents genetic speciation, while leading to bigger, better-connected societies with fancier technologies. Lastly, *Time* underlies our quest for objective, rational explanations for natural processes. The combination of knowledge and curiosity has driven us further than any other animal: we've developed the science to order the world and our place in it, becoming a connected global humanity.

It is the interweaving of these four threads that creates the extraordinary nature of us and explains how we operate as we do: why people who live in cities are more inventive, why religious people are less anxious, why Filipino storytellers have more sex, why migrants have a greater risk of schizophrenia, why Westerners see faces differently from East Asians. The human evolutionary triad – genes, environment and culture – are all implicated. For instance, the probability that any two of your friends are friends with each other – known as network transitivity – affects your individual fate, as well as the performance of the group.[1] But transitivity is influenced by environment – isolated villages have higher transitivity (everyone knows each other). On top of this, the number of friends you have is influenced by your genes.[2] The majority of all of this comes down to chance: who, where and when you were born is likely to be much more important than any choice you will ever make.

This is a great time to be exploring such fundamental questions about how we became such an extraordinary species. Exciting advances in population genetics, archaeology, paleontology, anthropology, psychology, ecology and sociology are beginning to reveal new insights into our history, fundamentally changing our understanding of how we developed as a species. For instance, the idea that a so-called behaviourally modern human emerged just 20 (or 40) thousand years ago, through some sort of cognitive or genetic revolution, is being challenged. The first individual human genome was sequenced in 2007, and since then, thousands of people have had their unique genetic history decoded and, in doing so, helped us understand our collective history – how we are related and how we relate to our closest human cousins.

Meanwhile, archaeologists using new dating techniques have made astonishing discoveries about our most ancient artworks and technologies, and paleontologists have shown the textbook tale of the simple ascent of man to have been anything but simple.

We are also entering a new era of collaboration: for the first time, many people from these famously protectionist research fields are beginning to talk to each other, upsetting well-established dogma but generating a wealth of data, insight and experience. This meeting of the natural sciences and the social sciences is starting to resolve this central paradox of why we are biologically so very similar and yet behaviourally so very different. We are looking at ourselves with new eyes, and recognizing the deep links that run through our biology, culture and environment.

As we will discover, human cultural evolution allows us to solve many of the same adaptive problems as genetic evolution, only faster and without speciation. We are continually making ourselves through this triad of genetic, environmental and cultural evolution; we are becoming an extraordinary species capable of directing our own destiny. It is this that has allowed us to expand our population size and geographical range, in turn accelerating our cultural evolution to greater complexity in a mutually reinforcing cycle.

Today, the size and connectedness of our populations have reached unprecedented levels. At the same time, we have produced a dramatic shift in Earth's environment, pushing the planet that formed us into an entirely new geological era known as the Anthropocene, the Age of Humans. The accumulation of our material changes alone – including roads, buildings and croplands – now weighs an estimated 30 trillion tons and allows us to live in an ultraconnected global population that's headed for 9 or 10 billion people.* Look around you: we are the intelligent designers of all you see. There is no part of Earth untouched by us – we're even littering space.

I'm going to take you on a journey to show how our uniquely human attributes changed us as a species – and how, in so doing, they reset our relationship with nature.

---

* Without this artificial habitat we would be reduced to a Stone Age population of no more than 10 million.

We are now, all of us, on the brink of something quite exceptional. The interplay of human culture, biology and environment is creating a new creature from our hypercooperative mass of humanity: we are becoming a superorganism. Let's call him *Homo omnis,* or Homni for short.

This is the story of our transcendence.

GENESIS

Every culture has its own creation myth to explain our origins, to make sense of the incredible unlikeliness of a talking ape that is curious enough to invent fantastical stories about how it came to be. The truth is no less remarkable.

Look up at the stars. You are not seeing them as they are now, but as they were millions of years ago. With your human eyes you are looking back in time, receiving light and imagery sent before our species existed, enjoying something perhaps long since extinguished.

We use history to look back at where we've come from. We also need science, because who we are is made from what we were before. Just as you might trace the origins of your cheek dimple to your great-grandmother, or find the basis for your country's politics in an ancient battle, so we must journey back though our ancestry to understand the origins of the structures, technologies and behaviours that drive our human world today.

Ultimately, this reveals our deep connections to our ancestor the sun. Our genesis is a story of physics, chemistry and biology that produced something that could control all three. Every one of us, everything on Earth, Earth itself, all the stars and galaxies in the universe, are deeply connected, and that connection goes back to a single point 13.8 billion years ago.

I

Conception

Fourteen billion years ago, the Big Bang created just enough of an excess of matter over antimatter for the existence of everything that we see in the universe today.

Entirety exploded out of something as contained as a quantum dot, and it has been expanding into glorious disorder ever since. Here on Earth, the only known living beings in the Universe attempt to do battle with entropy, create order out of chaos, building complex structures from energetic particles.

Energy generated matter, and that's made of atoms. Whether these crumbs make up a lump of iron, or an elephant's ear, or the scent of a rainforest depends on the number of protons at its heart: a hydrogen atom has just one proton, whereas lead has 82. But it is how the atoms transfer energy that determines much of the difference between hydrogen and lead (and their usefulness to us), and that is determined by their electrons, which spin outside the atom's nucleus and obey the strange rules of quantum mechanics.

The energetic exchanges made every time an electron moves within or between atoms are the basis for every reaction on Earth, from the replication of DNA to the laughter of a baby. It is the electrons embedded in our breakfast porridge that later provide us with the energy to chew our sandwich at lunch. These electron shifts allow atoms to combine chemically to form molecules, which are the building blocks of living cells, and so of us.

Around 90 per cent of all matter in the universe is hydrogen; another 5 per cent is helium, an unreactive atom with two protons. Both were produced in the instants after the Big Bang. As the stars shine, they fuse hydrogen atoms together into the heavier elements of our world,

3

including oxygen, carbon and nitrogen, which are extremely rare in the universe but make up most of the human body. And the violent drama that birthed the stuff of us also produced the elements we prize most. If you are wearing a piece of gold jewellery, know that you are likely to be wearing the celestial debris of a cataclysmic stellar collision so devastating that it literally shook the universe.

Gravity pulled together the interstellar clouds of hydrogen, helium and dust – the nebulae – with such force that their nuclei fused, releasing enormous amounts of energy and a new generation of stars. The star most important to our story, the sun – a nuclear reactor burning hydrogen in a cloud of cosmic dust – was born 4.6 billion years ago. Out of its dirty halo, a spinning clump of minerals coalesced: Earth, the third rock from the sun. Soon after, a massive asteroid crashed into our planet, shaved off a huge chunk – creating our moon – and knocked the world on to a tilted axis. That tilt gave us seasons and currents, and the moon's influence birthed our oceans' tides. Earth's position, the pull of Jupiter[1] and our orientation from the sun all played their part in creating a crucible for the greatest experiment in the universe.

Just one in 3 million of the molecules of Earth is water, but they are concentrated at the surface, and that makes all the difference. The ingredients for DNA, amino acids, rained down from comets, combined with the elements on Earth and kick-started life's incredible genesis in the planet's oceans some 4 billion years ago. At the nanoscale of atoms, where the masses involved are so small that the force of gravity becomes irrelevant, intermolecular forces such as electrostatic charges of attraction and repulsion dominate. One of the most surprising observations is that certain chemical processes become self-replicating. In this way, single molecules of DNA multiply, creating new life. Did the miracle happen just once or several times? We may never know for sure, but from one cell of self-copying magic evolved the incredible diversity of life that includes us, humans who have bitten the apple of knowledge and can now create nature itself.

Evolution has no aim and no direction – the ability to see, to walk, to fly, have variously arisen in creatures and been lost – but complexity takes time: billions of years of biological – and environmental –

evolution occurred before anything resembling us emerged. Initially, there was nothing to breathe, as the world's first atmosphere consisted of hydrogen and water vapour. It took around 2 billion years for the gas of life to pervade the air, courtesy of ancient blue-green algae, which used the energy from sunlight to make sugars from carbon dioxide, and in the process released oxygen as a waste product.

Photosynthesis and respiration, volcanic eruptions and tectonic movements, the tilt of the planet near or far from the sun – they all continually changed the balance of warming carbon dioxide and life-giving oxygen in the atmosphere, altering the climate as well as the chemistry and biology of the oceans. Over its first 3.5 billion years, the planet swung in and out of extreme glaciation. When the last ended, there was an explosion of complex multicellular life forms.

The emergence of life on Earth fundamentally changed the physics of the planet, and transformed it into a living, breathing system. When plants evolved, they sped up the slow breakdown of rocks with their roots, helping to erode the channels that would become our rivers. Photosynthesis imbued the Earth system with chemical energy, and when animals ate the plants, they modified this chemistry, releasing warming carbon dioxide and, with their death, contributing sedimentary layers to the original rock.

In return, the physical planet dictated Earth's biology, for life evolves in response to geological, physical, and chemical conditions. In the past 500 million years, there have been five mass extinctions triggered by supervolcanic eruptions, tectonic shifts, asteroid impacts and other enormous climate-changing events. After each, the survivors regrouped, proliferated and evolved as the Chinese whispers of random genetic mutations passed down the generations. The environment applies an evolutionary pressure on life, which selectively adapts – and it's been a two-way process: If plants became better at surviving in the desert (with genetic changes), they in turn changed the desert into less arid scrubland or dry forest. And this influenced which species (and which genes) could survive there.

There was no inevitability to our existence – to the existence of intelligent life – even if, looking back along our evolutionary route, it seems almost directed. Just an immeasurable rain of chance occurrences, big and small, splattering and splashing and, over the eons, trickling to unpredictable consequences: the delightful possibility of puzzle-solvers as different as an octopus and a human sharing space and time.

We can thank the heavens for our biggest evolutionary break. One day, in late June,[2] 66 million years ago, a meteorite so massive that it dwarfed Mount Everest, travelling at 14 kilometres per second (20 times faster than a bullet), plunged into the Yucatan Peninsula in present-day Mexico.[3] The impact was so extreme, so rapid, that the meteorite reached Earth still intact, exerting a pressure wave on the atmosphere in front of it that was so intense it began excavating the crater before the space rock even hit. On impact, the asteroid punched a 20-mile hole into the ground, deep enough to pierce the Earth's mantle, and sent shock waves across the planet that generated volcanic eruptions, earthquakes, landslides and blizzards of fire. What life survived the impact was mostly wiped out by the punishing global climate change that followed. Dinosaurs, which had dominated Earth for millions of years, disappeared; this ecological vacancy was filled by our mammal ancestors.

Some 10 million years later, rapid climate change turned the world humid, and tropical rainforests, palms and mangroves spread as far north as England and Canada, and all the way to New Zealand in the south. The Arctic Ocean was a balmy 20-something degrees Celsius but stagnant. Sea levels rose globally, and there were mass migrations and extinctions of animals and plants. Mammals diversified and the forerunners of many of today's common species emerged, including the first true primates. Then, around 20 million years ago, the Indian and Asian tectonic plates collided, buckling the land above so severely that the vast mountains of the Himalaya were created and the massive Tibetan Plateau uplifted, in a process that continues today. This new geography had a dramatic impact on the region's climate and biology: it divided the world's ape species into what would become Old and New World lineages, and produced new weather patterns, including the Southeast Asian monsoon. Meanwhile, volcanic activity beneath the Horn of Africa was tearing a great north–south rift down the eastern side of the continent, generating mountains with a raised valley between them, a process that fragmented the landscape and changed the climate. Evolutionary opportunities flourished in such dramatic environmental change.

We can trace our excellent colour vision to this time, when our foraging primate ancestor acquired the genetic mutation for an extra (third) eye cone cell, which enabled it to additionally see reds – most monkeys are limited to seeing in blues and greens. This helped with avoiding poisonous plants and distinguishing ripe fruits, which contain more calories, are easier to digest, and require less energy. Better nutrition allowed the growth of bigger brains; fruit-eating primates have 25 per cent more brain tissue than plant-eating primates.[4]*

Our ancestral habitat switch from forests to the savannah was another key turning point in our evolution. Its roots lie in a geological event 3 million years ago, when the drifting South American continent crashed into the North American continent at what is now Panama. This rerouted the ocean currents, dividing the Pacific Ocean to create on the east the Atlantic and Caribbean Sea. Warm waters from the tropics were pushed up toward the Arctic, where they cooled and descended down and south in a planet-spanning loop known as the Global Ocean Conveyor, which today dominates much of the world's climate. That led to the formation of the Gulf Stream, provided the moisture to freeze the Arctic, created a series of ice ages, and reset rainfall patterns, which dried East Africa and created new savannah landscapes.

Over the hundreds of thousands of years that our ancestors' bodies adapted to the savannah, climate change also reduced their former forest habitat. They had to spend many hours chewing roots and bulbs for protein, because there are no fruits on the savannah for much of the year, and increasingly relied on their social group for support. This particular orchestration of self-replicating chemicals had become a species ready to begin the process of self-domestication.

* The environmental role in seemingly unconnected biological traits, such as intelligence, is easy to overlook, but consider that plants evolved genetic adaptations that cause their fruits to redden and sweeten as they ripen (to attract animals that will spread their seeds). And primates evolved genetic adaptations in response to this environmental buffet. The environment shapes us and we shape it.

2

Birth

*The immense limestone monolith of Gibraltar rears up from the
southern tip of Europe – a stark, white geological totem visible from
Africa across a sliver of Mediterranean. At its base is a great teardrop-
shaped gash with a soaring interior: Gorham's Cave. What dramas
played out in this vast cathedral-like space? Whose lives . . . what
dynasties . . . lived, loved, worked, birthed, and died within these
ancient wave-sculpted walls? The cave was home for tens of thousands of years to our Neanderthal cousins – their last home on Earth.*

    *Let's go back 35,000 years: the continent is locked in a crippling
ice age, and those animals that could have left for warmer climes
amid plenty of local extinctions. In such harsh times, Gorham's
Cave is an idyllic spot. Sea levels are metres lower[1] and vast hunting
plains stretch far out to sea. Scouts higher up on the rock can easily
spot prey – or danger, like lions – and signal to those below. In
front of the cave's opening are fields of grassy dunes and spring-fed
lakes – wetlands that are home to birds and grazing deer. Further
around the peninsula are rich clam colonies and mounds of flint. The
line of neighbouring caves here has the highest concentration of
Neanderthals living anywhere.*

    *See the community busy with their daily chores: at the shore, chil-
dren gather driftwood; out on the plains, two women have ambushed
a vulture with beautiful black plumage and are carrying it home.
Let's follow them into the cave now. The main atrium, with its big
hearth, is bustling with activity – families are socializing, preparing
meals, working tools, making clothes. A broad-framed man in his
twenties, cloaked in tanned skins, is using a stone blade to sharpen
the end of a straight poplar branch. Wood shavings curl off his spear*

and he kicks them to the edge of the fire. By his side, a stocky red-haired woman is tapping open clams and skewering them with a sharpened bone – her sick aunt will be given this soft food first; they have already buried her child.

While the food is being cooked, an older man – perhaps he is a shaman – is fashioning a beautiful black feather cape and headdress from the vulture he's been given. These are people with rich interior lives, with time to think and create art. Deep inside the cave, past the little sleeping chambers with their individual protective fires, there is a special nook containing a deliberately carved rock engraving: a crosshatch of parallel lines. Its symbolic meaning will be lost in the befuddling layers of time, whereas creations made further north by other Neanderthals will prove easier to interpret: ochre animal paintings, handprints, necklaces of strung eagle talons and little ochre clamshell compacts.

They don't know it, how could they, but these remarkable people, who evolved outside of Africa with advanced culture and survival skills, are among the last of their kind. Within a single lifetime, drought transforms areas of thick hunting forest into unfamiliar grasslands. The few families that survive suffer more stillbirths, more weakening diseases. Perhaps they have already met the slighter human immigrants, who move in bigger caravans, establishing themselves in lands that were for thousands of years successfully occupied by Neanderthals. How vulnerable we are. What chance that it is I who sits here now and not she, the descendant of my long-extinct cousins?

If we are to answer the question 'What does it mean to be human?,' we might first ask what makes our way of living – our culture – different from that of other animals. Humans are exceptional: despite a growing raft of fascinating behaviours, no animal culture is anywhere near as complex or flexible as ours. Most animals rely on innate skills rather than learning from each other, and their culture isn't cumulative – unlike our technologies, the simple tools used by animals do not appear to have improved significantly over the past few million years.

Nevertheless, a variety of animals do exhibit some form of socially transmitted culture. Such species must be smart enough to learn a novel behaviour and social enough to transmit it. The most sophisticated tool users are our closest living relatives: chimpanzees – our last common ancestor lived 6 million years ago. Primatologists have identified 39 different traditions in chimpanzees across Africa (most populations use around 20), the most complex of which is nut-cracking.

For culture to become cumulative – that is, build on itself in a ratchet-like way with modifications that are selectively adopted and accumulate over time – the demands are far greater. A chimp can crack a nut by bashing it with a stone. Another chimp can learn this culture and it doesn't matter what sort of stone is used or how he bashes it, eventually the nut will probably crack. Developing nut-cracking further to make it more efficient would involve selecting a particular type or shape of stone, or perhaps even shaping the stone. In other words, it would mean adding steps, each of which has to be accurately remembered, in the right order, and demonstrated to another chimp, who would need to learn the steps and their order and also transmit them. Over time, modifications to the method can be made and new steps added until, eventually, the modern nutcracker evolves. As with genetic evolution, culture can only evolve with sufficiently accurate copying, which allows successful modifications – such as stone choice – to be maintained until they can be improved upon. Chimps are unable to do this, whereas we excel at it.

So when did this transition occur, the evolution of an animal with an extraordinary form of evolving culture?

Hard, isn't it, to look at a photograph of yourself as a child and reconcile that image with the adult in the mirror? There you are, the exact same person but for the passage of time and the experience of brain and body.

Looking back through time at those who lived thousands of generations ago requires greater effort in imagination and empathy. And yet those people were not so very different. They too were motivated, by the need for food and a safe home, to seek companionship and to come up with solutions to life's challenges whether social or technological. And they succeeded – some fleetingly; some, like *Homo erectus,* for more than a million years. We glimpse these long-gone cousins rarely but tangibly, finding a thighbone that once supported a purposeful run, or a skull that housed a thoughtful mind. More poignant than their bodies are the remnants of their humanity: tools worked by human hands or the marks they left on a wall, enduring testimony of the impulse to decorate.

Mostly, though, the millions of people that have lived before us have left no trace. They made clothes and tools from flesh and fibre, which have since rotted. Their bodies have been consumed and re-cycled back into the environment that bore them. We carry echoes of them in our DNA, in certain traits, in our human interactions, and of course we wonder about them, those pioneers of a different way of being, our cultural forebears.

From these clues, a raft of experts – among them paleontologists, anthropologists, geologists, climatologists – is trying to piece together a credible scene from a time when dozens of different human species lived on Earth. There is a famous drawing from 1965, the 'March of Progress' by Rudolph Zallinger, that illustrates human evolution as a parade of different hominids,* beginning with a primate ancestor and leading up to modern humans. It is usually interpreted as a linear progression in which each character is the direct descendant of the ancestor on his left, and pleasingly shows us out in front, winning the evolutionary race to be us. It is, as recent paleontological and genetic findings show, a cartoon whose only bearing on reality is the rel-atively recent arrival of modern humans. For a start, many of the different characters on the March were species that coexisted and probably interbred. Sex between different types of hominins resulting in genetic hybrids seems to have been commonplace, we are now dis-covering. Somewhere along this evolutionary transition, a special kind of culture arose, and tracing its emergence means casting back through our shared past for clues.

The earliest candidate is our distinctly modern, ancient human ancestor, *Homo erectus*, who emerged around 1.8 million years ago. By this time, hominin brain size had doubled from 600 to 1,300 cubic centimetres,[2] and these smart, prosocial people were able to remem-ber multi-step processes. Their tools were increasingly sophisticated,

---

* Hominid refers to humans and apes (living and extinct), whereas 'hominin' refers to modern and ancestral humans (those of the *Homo* genus, which first emerged around 2 million years ago).

unlike the simple tools made by earlier hominids, dating back 3 million years, which could be fashioned by an individual without input from anyone else. *H. erectus* was a remarkably successful fire-making, tool-using, sociable hunter who conquered continents ranging from Africa to Asia and the edges of Europe. They may well have had language and even made simple ocean-going crafts to travel to islands. Genetically, *H. erectus* was very diverse. Different populations spread geographically, intermingling and interbreeding with other related hominins over hundreds of thousands of years. Then, 1.2 million years ago, perhaps owing to climate change, *H. erectus* nearly went extinct, reducing worldwide to a population of just 18,500 breeding individuals.[3] For more than a million years, our ancestors were more endangered than chimps and gorillas are today, and it may be that this population bottleneck, which reduced diversity among hominins, propelled the evolution of our own species.

We don't know how many species of humans there have been, or how many different 'races'* of people, but the evidence suggests that around 500,000 years ago, Africans known as *Homo heidelbergensis* began to take advantage of fluctuating climate changes that regularly greened the continent, and spread into Europe and beyond. But by 300,000 years ago, migration into Europe had stopped, perhaps because a severe ice age had created an impenetrable desert across the Sahara, sealing off the Africans from the other tribes. This separation enabled genetic differences between the populations to evolve, eventually resulting in different races. It is from around this time that the very first evidence[4] for anatomically modern humans – *Homo sapiens* – appears in Africa, where they would develop their cultures and intermingle and breed with other (now extinct) African races, such as the recently discovered *Homo naledi*. Those hominin populations that had left Africa adapted to the cooler European north,

---

* I describe the different cousin species of human, such as *H. sapiens, Neanderthals, Denisovans*, as different races, because they were genetically similar enough to have interbred with each other successfully, producing fertile offspring. While individual hominin groups may have lacked diversity – they appear to have been quite inbred – their collective diversity was great. For most of our history *sapiens* have existed among other races. Today, there is only one genetically distinct race: ours.

eventually emerging as Neanderthals, Denisovans and others whom we can now only glimpse with genetics.

By the time the first few families of modern humans made it out of Africa,[5] around 80,000 years ago, Neanderthals were thriving from Siberia to southern Spain. Today we find their ghosts living on in our genes – for it seems that wherever we encountered other humans, we bred with them.[6] Everyone alive today of European descent – including me – has some Neanderthal DNA in their genetic makeup, and across the population as much as 20 per cent of the Neanderthal genome is still being passed on, presumably because it has helped us survive in Europe.* Other archaic human races have also left a genetic legacy in modern populations. Indigenous Australians carry genes from Denisovans, a race about whom we know very little, while other, yet-to-be-identified archaic races have influenced the genes of other human populations across the world, including as recently as 20,000 years ago in Africa.[7] Perhaps it is our lascivious nature, which enabled us to collect so many usefully adapted genes from the various hominins we encountered, that helped our ancestors succeed as they spread across the world's environments.

Imagine that time in our history, where people could encounter those of a truly different race – other types of human trying the same cultural experiment. How fragile we all were. What a risk our evolution took when it put all of our survival eggs in that one basket of culture, pitting us against fearsome beasts and cruel climates. We were physically so unprepared for hostile conditions that for most of human history, our survival has been touch and go. Just 74,000 years ago, for instance, a supervolcanic eruption at Toba in Indonesia nearly wiped us all out, and our ancestors' population shrank to a few thousand. Today, although there are several species of ape, only one human species has survived.

---

* Many of the genes we have inherited from Neanderthals are associated with keratin, the protein in skin and hair. These visible variants may have been sexually appealing to our ancestors (Neanderthals were redheads), or perhaps their genes for tougher skin offered some advantage to the African migrants in the colder, darker European environment. Some Neanderthal genes are now problematic – a gene that may once have helped people to cope with food scarcity now leaves Europeans more prone to diseases like type 2 diabetes.

For us alone, the cultural gamble paid off; all of our similarly talented cousins went extinct, leaving a mere fragmentary record of their hundreds of thousands of years on the planet. So, if we are to pin our astonishing global success on our culture, we should recognize that our glorious ascent wasn't a given. Nothing makes that clearer than the tragic failed attempt of our extinct Neanderthal cousins – they too were cultured and, compared to us, stronger, bigger-brained and better used to surviving freezing conditions. So why did only we succeed?

Partly it was down to luck. The climate changed in a way that favoured savannah hunters. We may have been carrying diseases to which the Europeans had no immunity. Most importantly, though, Neanderthals were very inbred[8] by the time we encountered them, and their populations were roughly one-tenth the size of contemporaneous sapiens'. Geneticists estimate that Neanderthals had an at least 40 per cent lower evolutionary fitness, the measure of how well a species is able to survive and reproduce, than the modern humans at the time of contact. This reduced their relative population and genetic diversity still further. In computer simulations of our Paleolithic interactions with Neanderthals, using estimates of population sizes, migration patterns, and ecological factors, Neanderthals go extinct or become completely assimilated[9] within 12,000 years of our species' arrival.

Evolutionary success is measured ultimately in numbers – and there were simply more of us, trickling into Europe. But why? Had we evolved a greater intelligence that led us to outsmart our cousin species, as is commonly believed? It is possible, but all the evidence points to us being very like the Neanderthals in terms of brain size and the tools we used. Nevertheless, there was clearly something about *our* biology or culture that had enabled our numbers to flourish and had given our populations greater resilience in an exceptionally harsh environment, when as much as one-third of the world's land was covered in ice.

I think that the size and diversity of our gene pool holds a clue to a bigger truth about the size and diversity of our culture. We had better-connected, bigger populations that collectively held more cultural knowledge from which our ancestors could draw. Our ancestors may have been slightly better at socializing, slightly better at learning from each other and slightly more curious about the world. I think it is telling that even though Neanderthals survived for hundreds of thousands of years, they never left their native landmass, whereas our own ancestors were already global explorers. As the fossil record testifies for countless species, global distribution offers the best survival chance during catastrophic events.

As we will see throughout this book, our success as a species has long been tied to the changing nature of our environment and to the size and shape of our societies. These are mutually connected. Rapid climate change and population expansions or contractions result in bursts in human innovation and cultural activity – or their opposite. Through it all, we experimented and learned and we taught each other the tricks of survival. We spread across the globe, inhabiting all the various geographic niches of our world, and our genes obligingly adapted. We were born a species entirely determined by our planet. As our culture developed, we began to modify our Earthly nest and control our fertility, till we became the only species to determine its own destiny.

Let's investigate this transformation through four key elements, beginning with the spark that fuelled our cultural evolution.

FIRE

All life needs energy to function and gets it from food or, in the case of plants, the sun. Humans, though, are also able to harness wild forms of energy, and it has been transformative. We are who we are because we outsource our energy costs, enabling us to escape our environmental limitations and extend our physical capabilities. So how did we kindle a new relationship between environment, biology and culture?

3

Landscaping

December in Queensland, Australia, is hot. On the drive down the Pacific Highway, past sugarcane fields and through open bushland, the monotonous hum of cicadas resonates in tune with a heat haze liquidizing the asphalt ahead of my sticky rolling tyres. Hot air blows hard across flatlands. The cloying sweet odour of cane fields is diluted, replaced by the heady skunk of green kamala and sharp pungent eucalyptus. Here and there, scrubby bushes become trees. Lizards, snakes, birds – mostly dead – flash past me. Occasionally the road curves but soon restraightens. Hour upon hour of tarmac conveying me south at a steady 80 kilometres per hour.

Before I notice it, the arid green expanses either side of the road have blackened. I register the change with mild interest, noting also a new quiet. I drive farther. Now there is smoke, a grey layer hugging the burnt ground. The road ahead becomes bird-filled. Black corvids and raptors float and gather along the highway, looking for creatures fleeing the scalded vegetation. Black on black on black. Further up, the smoke becomes thicker, filling the windscreen. I have entered an eerie, alien world devoid of colour, unnaturally dark. The sulphurous stench of burning increases. Bright pools of luminous gold flash out of the gloom – smouldering fires that become more frequent until the edge of the road becomes a dancing river of fire.

I start to worry that this might be dangerous. Cocooned in my car, I see the same view through the windscreen and rear-view mirror: flashing flames against thick, disorientating smoke. I slow my speed.

To left and right the many pools of isolated fires become more frequent, getting closer together and bigger. I can hear the fire now, a roaring, crackling, hissing dragon. Suddenly I am surrounded by high walls of flames, towering above me, licking and sucking the air around my car. The heat warps the air, bending the light. The roar is deafening, a gale of fire. Smoke seeps through my closed windows and I panic.

Time slows down, seconds bleeding apart as the noise dampens, my vision tunnels, and my arms lock onto the steering wheel. I slam my foot onto the accelerator and within a couple of minutes I've outpaced the fire. Behind me, smoke billows high above the retreating flames; ahead there is colour. I lower the windows, breathe the antiseptic camphor of eucalyptus, let my eyes bathe in greens and blues and listen to bird squawks outdo my thudding heartbeat.

In our tame, human-made world, when it is rare for most of us to encounter any threat from the wild, fire maintains terrifying power. Fire ravages landscapes and property, and is still a major killer. My few minutes inside that burning hellscape left a deep, visceral impression. Fire is primeval.

Yet there was a time before fire. A time when Earth, formed as it was in the sun's explosive furnace, could not itself sustain flames.

For the first billion or so years of Earth's history, there were no fires because there was nothing to burn and no oxygen to burn it. It took the evolution of photosynthetic bacteria, followed much later by the growth of the world's first forests, for there to be the ingredients of fire. Life itself had to generate the environmental conditions for its own destruction.

Fire is chemistry made visible: a marriage of oxygen and fuel in an exuberance of heat and light. This is the same basic reaction that sustains all life – it is how we get energy from our food – but in living cells, it is called metabolism and is a slow step-wise process, whereas fiery combustion is lightning-fast and intensely energetic. Our ancestors learned to seize this raw energy, harness and subjugate it for their own use. Humans were the first creatures to use fire to remake the environment that made them, expanding their ecological niche and changing forever the dynamic between ecology and random 'acts of god'.

When humans began deliberately accessing resources of energy beyond their own muscle power, they transcended the realm of biological life and entered a new state of being. This bounty of extracorporeal energy enabled an entirely new form of selective adaptation: cumulative cultural evolution, which would come to define our species. As our ancestors developed the cultural ability to exploit external forms of energy, so this nurtured and reinforced the cognitive and social conditions for their cultural growth – our brains grew and we became more social, cooperative and better at learning from each other. Energy supercharged our species, and the quest for energy efficiency would drive our cultural evolution, change our genes and make cyborgs of us all.

It started, millions of years ago, with wildfire.

As a fire consumes forested areas, expunging habitats and food sources, it also opens up areas to new plant growth – to grasses, for instance – and recalibrates the hierarchy of survival for other plants and animals. Large herbivores become far more numerous in open savannahs, and so do the carnivores that hunt them.

The power of fire to change the food density of a landscape would not have gone unnoticed by our ancestors, and at some point in our evolution we began to exploit it. Our early forest-dwelling ancestors would have observed, as birds do, that fires leave in their wake exposed forest creatures that are easy pickings. As bipedalism increased, these primarily vegetarian hominids gained better access to open landscapes, and their taste for meat grew. In Ethiopia, there's evidence from 3.4 million years ago that foraging *Australopithecus* people feasted on cow- and goat-sized animals.[1] The raw meat had been butchered with stone tools and the bones smashed for their marrow by these primitive ancestors, despite the fact that they had not yet acquired the anatomical adaptations to their teeth and jaw for regular meat eating.

Chewing and digesting raw meat is tough, whereas scavenging cooked carcasses (and plants) is a tastier, more hygienic and more efficient way of getting calories. That's because fire chemically changes food, making it easier to digest. People who ate cooked meals would have been healthier and more likely to survive long enough to pass on their genes and food-sourcing habits to others, making fire-based buffets an increasingly important part of our ancestors' diet. The distinctive smoke plumes of a bush fire could have attracted groups from great distances.

In time, our ancestors learned to capture wild flames to produce their own fires. This was a giant step in our relationship with fire, so it is remarkable to note that some populations of Australian raptors, including black kites, also have a fire-spreading culture. Known to Aboriginal people as 'firehawks', the birds pick up flaming twigs from wildfires and then deliberately start fires elsewhere, in order to flush out prey from grasses. It is easy to picture our smart ancestors doing the same, millions of years ago, and then carrying embers from camp to camp. Good, dependable social networks would have been essential to maintain these legacy fires sustainably over time and across different locations. So, as we became ever more reliant on fire, we thus became more reliant on each other.

Fire was a security blanket. Whereas our earliest human ancestors had bedded down in tree nests for safety, fire protected their descendants from predators and the cold, allowing them to sleep in open savannahs. In other words fire culture was adapting our species' habitat for their survival; as fire made our world safer, we altered the environmental selection pressures acting on our genes. We were not the first animal to alter its environment, of course, but most other creatures do this instinctually, meaning they are genetically programmed to modify their environment in a species-specific way. Beavers may construct dams and ants make complex mounds, but never vice versa. Humans, by contrast, are not preprogrammed to any specific environmental modification, but we are exceptionally creative,[2] and over time our ancestors' genes evolved in response to this new, culturally determined environment. We became fully bipedal, losing climbing feet in favour of flat feet adapted for better running – this is likely only to have been feasible if night-time safety was assured with fire protection.

Next came fire making. It is a skill we have to learn, and on which we are entirely dependent. So essential was this fire-conjuring trick that every culture has adorned its origins in elaborate myth. For the Ancient Greeks, fire was the greatest gift to mankind, stolen from the gods by Prometheus. For a theft of this magnitude, Prometheus was chained to a rock for eternity while his liver was picked at daily by an eagle. The Arctic Yukon people tell that Crow stole fire from a volcano in the middle of the water; whereas for the Ekoi people of Nigeria, fire was stolen from the creator god, Obassi Osaw, by a small boy, who taught the people fire making and was punished for his theft with lameness.

I imagine a more prosaic beginning. The act of working stone tools, chipping one with another, would have produced sparks. From there, it is not a huge leap to imagine our ancestors kindling a fire. Nevertheless, as far as we know, it's a leap that only hominins made. The earliest evidence we've found of human fires – they do not preserve well – are from 1.5-million-year-old sites in the rich archaeology of Turkana in East Africa's Rift Valley.[3]

Fire making can be as simple as rubbing a stick into a groove on another piece of wood. I've done it myself after a memorable hunt with a group of Hadzabi hunter-gatherers in Tanzania. First I was shown how to cut a notch out of a broad, flat bit of wood – the 'hearth' – which I held fast between my feet, whilst seated on the ground. Then I was given a smooth, straightish stick similar to a pencil. With the pointed end of the stick standing hard into the groove, I twirled it back and forth between my palms, creating friction in the groove. It took a couple of minutes for smoke to rise. Scraps of tinder – dried shavings of an oily bark – were put in the groove, and my teacher took the wood between his cupped palms and blew life into the fire.

The simplicity is deceptive, though – had I not been shown how, I doubt I'd have been able to invent this fire-making method. For a start, the choice of stick and hearth woods, and where to find them, were important yet unobvious. One of the Hadzabi men had tied string around the fire-starting stick to twist it back and forth, creating an efficient type of drill that saved his palms. It was a modification he had been taught and would teach to others. Evidence found at several Neanderthal camps in France suggests that one particularly sophisticated way of starting fires involved the lustrous mineral pyrolusite (manganese dioxide),[4] which lowers the temperature required for combustion. Archaeologists found large stores of the small black blocks and believe they were powdered and mixed with a tinder fungus to produce fire on demand, just as we use matches today. Whichever method a group uses, it is passed from one generation to another, the information[5] as precious as the materials used.

This simple spark demonstrates a key difference between how hominins operate versus all other animals. While primate cultural practices are simple and, on their own, easy enough for an intelligent individual to innovate by themselves, fire making is multi-step and complex. By the time of *H.erectus*, more than a million years ago, the collective cultural tool kit contained so many different and complex skills, from fire making to tool production, that it would have been impossible for an individual to come up with them all in a single life-time. Instead, they were learning from each other, practising and remembering the details in such a way that their cultural knowledge had become cumulative – human culture was building and evolving, and our ancestors' brains had evolved primed for cultural learning.

So was it our bigger, smarter brains that enabled fire making, or was it fire making that enabled our bigger brains? Well, it was both; a mutually reinforcing evolutionary process that took hundreds of thousands of years of feedback as our genes, culture and environment all adapted. As the Greeks realized, fire gives us godlike power over nature and hominins became landscapers of their environment, using this energy to improve grazing for the herbivores they fed on, creating ecosystems* that suited their needs and improved their survival.

In other words, our ancestors constructed environmental conditions that favoured their reliance on culture – the more they were able to control and regulate their environment (and that of their children), the greater the advantage of transmitting cultural information across generations. We were making ourselves.

One of the consequences of our landscaping and move to the savannahs was that it made it easier to hunt bigger animals containing more calorific fat and meat. The first evidence of human hunting is about 2 million years old[6] and marks an important cultural development that changed our ancestors' anatomy and behaviour.

For millions of years, hominids had been primarily vegetarian foragers, but as their culture and environment changed to support a meatier diet, so their bodies also adapted. By the time of our *Homo* ancestors we had evolved into endurance hunters, running on sprung,

---

* Such an extensive domestication process would lead eventually to cultivated species of plant and animal, the creation of artificial life and the invention of landscapes so unnatural that they may be inhabited only through virtual reality headsets.

high-arched feet, with narrowed hips and pelvises, muscled buttocks and an S-shaped spine to carry our newly flat-faced heads. Our torso and arms had lengthened to counterbalance our stride and provide stability. We had also developed a novel ability to throw projectiles. Although some primates occasionally throw objects, only humans can routinely hurl stones or spears with both speed and accuracy, the result of catapult-like adaptations to the shoulder and torso, which anatomists estimate occurred some 2 million years ago.[7]

We had also shed our body hair and dramatically increased our number of sweat glands, which enabled us to maintain a safe body temperature through evaporative cooling while running in the heat of the tropical sun. It may be that a single gene caused our hairlessness and also led humans to have the highest density of sweat glands[8] of any primate, able daily to produce litres of cooling sweat. At around the same time, a gene for dark skin appears in our ancestry, which protects us from ultraviolet sunlight. Our genes responded to our cultural behaviours, enabling us to outlast any animal on the savannah, able to run beyond its endurance or until it was in our missile range.

This suite of anatomical changes – which improved our diet and was therefore highly adaptive (in other words, they were evolutionary changes that improved our survival in the environment) – was coupled with cognitive, cultural and social transformation. It is clear our genetic evolution had shifted trajectory: humans don't look like the other savannah hunters, with our puny bodies, our lack of sharp teeth and claws, but our cultural and anatomical changes have made us the deadliest creatures alive. Culturally, our hunting tools and weapons were, even 2 million years ago, more varied than the sticks used by chimps or the sponges wielded by dolphins to flush out their respective prey, and they were deliberately manufactured. Unlike other animals, our ancestors used a selection of tools and invariably processed their kills afterward, making use of bones, horns and skins for other purposes. Using a specific tool for a specific job is more efficient than maintaining vast biological reserves of less specialized muscle energy. Hunting became a learned cultural adaptation whose multi-step practices would evolve over thousands of generations to the global mechanized meat-producing industry of today.[9]

*

In turn, hunting fundamentally changed our society. It introduced divisions of labour between hunters and gatherers, and longer settlement periods. At the same time, our dependency on the campfire became mutual: the fire became another hungry member of the group, needing constant attention and refuelling, which meant frequent, increasingly long, and costly journeys in search of firewood at a time when people were living a hand-to-mouth existence. To accommodate these extra labour costs and make hunting more energy efficient, human societies responded by forming larger, multi-generational bands.

In other words, hunting made us social. A hunt could require three or four people working in concert, or a larger group to take down a big prey like an elephant. To be successful as a team, each person would have to be able to anticipate the actions of each of the other hunters as well as any other predators out there, by imagining the thoughts and perspectives of other minds. It took hours of persistence, skill, careful observation and strategy. We learned to recognize and follow animal tracks and to understand their behaviours. The hunt would be deliberate and planned: we would mentally conjure a future scene – one in which it is hours later and we are thirsty – and communicate this idea with our group. Water would be carried in a skin or bladder. We can only outlast stronger animals because we have our own supply of water to replace our sweat, and because we can train our minds for endurance – we have mental strategies to urge each other on, to keep going even after we are physically exhausted. We can override our biological impulses and urges and push on through 'the wall'. When the human body is under stress from physical exertion or starvation, blood flow is prioritized to the brain over muscles[10]: somewhere in our evolution, quick thinking became more important than fast moving.

Hunting was socially and mentally complex as well as being physically demanding and risky, but the pay-off of more calories outweighed these energy costs, and the mutually reinforcing evolutionary process propelled us on.

The intellectual rigours of a cooperative hunt require a bigger frontal cortex, the part of the brain that deals with social behaviour, decision making and problem solving. That is why lions, the only big cat to hunt in groups, have the most highly developed frontal cortex – lionesses, which spend more time in groups and do the majority of the hunting, have the largest frontal cortex.[11] Studies have also found that when dolphins hunt cooperatively with human fishermen, those co-operating best with fishermen are the most social with each other – their closeness increased the odds that one dolphin would learn the co-operative hunting technique from its peers.[12] Novel behaviours can only spread if animals are social and therefore have the chance to copy from each other. Humans have used the sociability of a range of animals to acquire their own calories more efficiently, something that probably preceded domestication. For instance, several sub-Saharan commun-ities rely on a partnership with the small honey bird, which responds to their calls and guides them to beehives. There, the humans can smoke out the bees, allowing both partners to retrieve honey, which accounts for as much as 15 per cent of the calories consumed by some hunter-gatherer groups.

We rely most on each other, though. Unlike other primates, in-dividual humans don't hunt for themselves alone, and food is brought back to the group to be distributed – there is evidence of hominins carrying food back to a home base 2 million years ago. This enables greater hunting efficiencies through specialization – the best spear maker might not be the best spear hunter but the group benefits from both skills and can acquire more food for all. Cooperation and food sharing increase a group's strength, and lead to a greater diversity of complex skills. Although hunters are in their physical prime in their twenties, individual hunting prowess does not peak until age 40, be-cause, for humans, success depends more on know-how and refined skills, which take time to learn.[13] In hunter-gatherer societies, most hunters do not produce enough calories to feed themselves, let alone others, until they are aged around 18. By contrast, chimpanzees, which are also hunter-gatherers, can sustain themselves immediately after infancy, around the age of five. Being even partially dependent on others for your meals risks hunger if you're cast out of the group or there's not much to go around, but group reliance, and the cooperation it depends on, has important survival benefits for the group as well as the individual, and for us it won out over self-reliance.

The better we got at relying on the energy efficiencies of group life – tending fires, strategically burning and hunting cooperatively – the more food we had as individuals, the better we survived and the more likely that our genes were passed on. Socializing takes energy and time, but it increases survival, so it triggers biologically evolved mechanisms that reward it. All primates spend several hours of their day grooming each other as a physical method of socializing; this creates and maintains lasting bonds that secure their place in the group's social strata. Grooming releases endorphins – natural opiates – in the animals, so it feels good, encouraging more social behaviour. We, too, get pleasure from socializing. A neural circuitry[14] in which social contact is 'rewarded' with an oxytocin or dopamine hit mediates our behaviour so that we seek such experiences again. Group activities, especially when done in synchrony, like music making or dancing, release these same drugs in our brains, programming us to seek another hit. Equally, social rejection really hurts – it elicits the same responses in the brain as physical pain. However, rather than spending the precious daytime grooming each other, our human ancestors extended the day with firelight, giving them time to socialize after dark. Adult humans have an exceptionally long waking day of around 16 hours or more, compared with around eight hours for most mammals, and early evening is the beginning of 'social' time for cultures across the world.

This new, culturally evolved creature that hunted strategically with an arsenal of tools and fire had a dramatic impact on the landscape. East Africa today has six large carnivores: the lion, leopard, cheetah, spotted hyena, striped hyena and wild dog; before 2 million years ago, there were as many as 18, including bears and civets, sabre-toothed cats and bear-sized otters. Their numbers began dropping precipitously as our ancestors began hunting,[15] and it was the same story wherever we went. By the end of the Pleistocene, some 11,000 years ago, around 5 million humans had wiped out an estimated 1 billion large animals. Even if we didn't hunt all these big beasts to extinction, we would have been in direct competition with them for prey – targeting the same prey or driving them from their kills during our scavenging. Being omnivores, we could always fall back on foraging during lean times, unlike big cats. The loss of so many top predators transformed the East African ecosystem in a so-called trophic cascade, producing an explosion in small mammals and herbivores, and further reducing tree coverage. Taking their place was the most successful predator the planet has known. Today, most large animals are afraid of projectiles – an adaptation to us alone.

Our evolutionary triad had multiple effects on the ecosystems we inhabited, altering the evolutionary trajectories of many plants and animals. This, in turn, altered our own evolutionary journey. As the herbivore population diminished and became fearful of humans, so spear hunting became harder for the same calories. Those who were better at spear hunting had a selective advantage and so, over generations, we became better at such skills both anatomically (physically better hunters pass on their genes) and culturally, because if we grow up in a cultural environment in which everyone is practising such techniques, we become more skilled.

Fire-mastery was key to our human toolbox: it didn't just allow us to change our landscape, it also freed us from the evolutionarily determined tropical niche that still binds our primate cousins. We could follow migrating herds of food, set up camp where we wanted and deliberately change ecosystems that weren't suitable habitats. *Homo erectus* pioneered the human spread across the globe, living from the tropics to the coldest latitudes. Hundreds of thousands of years later, tribes of *Homo sapiens* undertook a similar exodus, relying on a network of aquifer-fed springs during a rare wet spell, to venture out of Africa. It was a slow spread: we moved a kilometre a year, on average, to make our way into the Middle East and then farther eastward according to the timescale indicated by archaeological and ancient DNA evidence.

From the Middle East, some *sapiens* made it all the way to the great expanse of Australia (which was connected at the time to New Guinea). It is likely that smoke from large bushfires attracted people to undertake that first audacious human sea voyage some 60,000 years ago – an intrepid migration across 100 kilometres of open ocean.[16] Smoke meant fire, which meant land covered in vegetation with all the possibilities of plenty and peace (far from competing tribes), which every migrant dreams of. It was an extraordinary voyage by a remarkably sophisticated species, and worth it: on arriving, the first Australians found a land of giant marsupials, birds and reptiles on a vast unpeopled territory.

*

Over time, fire-domesticated landscapes around the world have become dependent on us for regular burnings. In Australia, the practice of 'firestick' farming dramatically altered the ecology of the continent, creating a mosaic of dry forest and savannah, increasing the numbers of kangaroo and other grazing marsupials, as well as promoting the growth of edible fruits, flowers and plants, such as bush potatoes. This sort of land management ensured the survival of fire-tolerant plants and reduced the fuel load in any one area, so the enormous blazes that are now so frequent in Australia were relatively contained. Across Africa, savannah cultures typically burn an area half the size of the continental United States every year, setting fires to keep grazing lands productive and free of shrubs. But fires are declining as they and those in the Eurasian steppes and South America move from nomadic to agricultural lifestyles. Between 1998 and 2015, global fires shrank by 24 per cent a year, or 700,000 km^2, reducing habitats for some of the most endangered carnivores. Nature created cultural beings that subjugated and enslaved it until it is now dependent on us for its continued existence. Today, most of the world's fires are made by us.

4

Brain Building

On Sunday, 11 March 2018, Emily Dial, an experienced midwife, scrubbed up as usual for a routine caesarian section. Once prepped, she joined the rest of the medical team in the delivery room at Frankfort Regional Medical Center in Kentucky for the brief procedural talk-through, while they all donned surgical gloves. Then, she climbed onto the operating table, lay on her back and raised her gown.

It was the anaesthetist who numbed her so his colleague could slice through her belly's layers of fibrous tissue, but it was Dial who delivered the baby.

Her hands were guided by the doctor into the incision he'd made. The room fell silent but for the steady beeping of medical equipment. Carefully, Dial felt around the baby's emerging head, then, with an expert grip, she turned and pulled her daughter's slippery body out of herself. As she carried the pink, rumpled newborn up to her chest, the room erupted in applause and the unmistakable cries of an infant. The midwife had delivered her own baby.

However remarkable the procedure,[1] humans need help to give birth because our babies have big heads for such a slight maternal frame. That's because most of the energy efficiencies that our anatomy has evolved, in response to the cultural and environmental changes we have made, prioritize our brain over the rest of our body. Humans are puny compared to a chimpanzee, but our brainpower far exceeds theirs. By harnessing fire, our brains could grow beyond the limits imposed by biology. It left us unable to give birth alone, but smart and social enough to survive anyway.

We've seen how our adoption of fire enabled us to change our environment, and how that affected our biology and culture. Now let's see how fire enabled us to grow our exceptional brains. Much of our distinctiveness as a species is down to our brain size – getting there involved an intricate dance between culture, biology and the incorruptible laws of physics.

Generally, as animals get bigger, their brains get bigger – and this correlates with increased intelligence, sociality and culture. Dolphins, for example, resemble humans in several of their behaviours and cultural activities: they play, babysit each other's offspring, hunt cooperatively, use individual names (signature whistles) and teach one another. But such sociability and cultural richness is only seen in those with the biggest brains.[2] When animals have bigger brains than would be expected for their body size, they turn out to be the smartest. Chimpanzees have got brains twice as big as the average animal of that size. Our brains, however, are the largest of any primate relative to body size: seven times bigger than expected, and more than three times bigger than a chimp's.[3]

Sociability requires a bigger brain but it is also a consequence of a bigger brain. Over generations, as our ancestors relied more on their smarts for survival, they became bigger-brained and more social because people with those characteristics were more likely to survive long enough to bring up children. Geneticists recently discovered three nearly identical copies of a gene that occurs only in humans and are involved in brain development, which they believe supercharged brain expansion in our ancestors.[4] The first duplication occurred 3 to 4 million years ago, around the time of the first crafted stone tools. The gene then duplicated twice more to create the versions that modern humans carry. Across mammals, the genes shown to have the fewest changes over time – in other words, the most crucial ones – are those involved in the brain. Except in us: in humans, over 90 per cent of the genes expressed in the brain have been up-regulated over the past 2 million years, making their effects more potent.[5]

Our superior intelligence isn't only down to brain size; the number and relevance of the neurons we pack in is important too. We have an exaggeratedly large cerebral cortex, for instance, which is the part of the brain involved in higher cognitive function such as perception, memory, language and consciousness. It is a wrinkled layer of nerve tissue, just a few millimetres thick, but if it were stretched out, our cortex would cover an area of roughly four A4 sheets of paper, whereas a chimpanzee's cortex would fit on just one sheet, a monkey's on a postcard and a rat's on a stamp. The thickness of the cortex and size of key parts is also important – people with a thinner cortex develop a lower IQ,[6] whereas those with a bigger prefrontal cortex tend to have more friends.[7] It is likely that our ancestors increasingly selected partners who were cleverer and more sociable, rather than beefier and more aggressive. We domesticated ourselves.

But our expanding brains came with significant risks. The very same selection pressures that had caused our brains to grow – a successful hunting culture – had also caused adaptive changes in the rest of our bodies: we had evolved narrow hips and a small pelvis as highly efficient, bipedal runners. Consider that female chimpanzees, who stand just over half the height of female humans, have roughly the same-sized birth canal, yet the brain of a newborn chimp (roughly 155 cc) is less than half that of a human newborn. Getting a giant head through a narrow birth canal without mother or baby dying is a challenge.

While death of the offspring is always an undesirable outcome for a species, maternal death is often not. For many animals, the mother dies, is eaten or disappears soon after birth, but this is not the case with mammals and certainly not with primates. Cultural species, which rely less on instinct and more on learned skills and behaviour, require long-term parental attention and care. Maternal survival would be vital for the continued survival of our species.

Solving our obstetric dilemma required the evolution of adaptive changes to our sociality and to our anatomy, including engineering a temporary reduction in the baby's head size. This was achieved through delayed fusion of the fetal skull, which at birth is still in six separate bone plates that can overlap and move, enabling the head to deform to pass through the birth canal. Babies are also born with brains less than one-third (28 per cent) of their adult size, while a newborn chimp brain is 40 per cent of its adult size. Our babies are born so immature that the first three months postpartum are often referred to as the 'fourth trimester'. To navigate the 'hunter's' pelvis, human babies also evolved to perform risky rotations. Ape infants fit easily through their mothers' relatively roomy pelvises and typically do not turn. They are born with their heads facing upward, toward the mother, from where they can be pulled easily to the nipple. Our 3-million-year-old ancestor Lucy (*Australopithecus afarensis*), who was also bipedal, gave birth to an infant who would have had to turn once (45 degrees), so that it emerged facing sideways into the mother's leg. Our babies, though, are born after a double turn that risks wrapping the umbilical cord around the neck and leaves them facing toward the mother's tailbone.

Despite all the adaptive anatomical modifications of a smaller immature brain and collapsed skull, humans in every culture across the globe need assistance to give birth safely; our hypersociality, which required such a large brain, evolved in step with the need for assisted birthing. Friendships and collaborations between women would have been key to this, as they are today, and key to the strength and survival of the group as a whole.

Even after the birth, human mothers need assistance to keep their newborn alive. Until I had my own children, I assumed breastfeeding was innate and simple – after all, as the defining feature of the mammalian taxa, it must be as obvious as breathing, right? I was surprised to learn that not only did my baby have no clue, neither did I. The baby's gape, position on the breast and timing had to be taught, and took time and practice. It was a week or so until I could breastfeed my infant as naturally and easily as a nursing chimpanzee. In every culture, women are taught how to breastfeed after birth, and infants whose mothers have difficulty are often nursed by other women in their family or community or, more recently, fed from specially produced bottles of formula created to match the nutritional make-up of human milk.

For birth and breastfeeding – the most crucial events in the transmission of our genes and survival of our species – to have become such difficult tasks that they need to be taught, cannot be done alone, and carry a high risk of mortality for both mother and infant, means that the payoff – a bigger brain, extreme sociability and cultural know-how – must have been worth it in terms of our evolution. And, as with so many of our evolutionary changes, they arose in concert with fire-mastery. It is unthinkable that difficult childbirth would have routinely taken place without the protection of fire – living on the plains meant exposure, and human babies cannot jump up and run for safety like baby gazelles. The anatomical changes towards very large brains, and the birthing difficulties that resulted, must have taken place after our ancestors learned to control fire.

Our cultural adaptation for cooperation was so successful that we began to rely on it for childcare, too. Most mammals can stand and follow their mother soon after birth; human babies cannot even roll over. The delayed skull fusion allows for two full years of brain growth before the skull hardens, which means relying on sufficiently protective social care of the soft-skulled infant during this time. Human brains expand much more dramatically than chimpanzee brains during the first few years of life, mostly driven by explosive growth in the connections between brain cells, an expansion of white matter. While absolute brain size is important to intelligence, much of our cultural learning is achieved by forming new connections *between* cells rather than new cells. Human brain growth and development continues until at least age 30,[8] enabling our extraordinary neuroplasty – our brain's ability to reorganize and grow new neural connections throughout life in response to new information, environmental conditions, or injury – and extended capacity for learning. So even after weaning and learning to walk, children need investment of time and resources from their parents and community to survive childhood and become socialized adults with the skills to negotiate their place in the tribe.

Our babies take longer to gestate, and need greater and longer care post-birth, and yet the sibling gap for humans is briefer than for apes. Human mothers can produce babies at yearly intervals, although more commonly two to four years later. Chimpanzees, by contrast, have five-year intervals. This difference alone means that human populations can increase faster, social groups become bigger and more complex and culture can advance more swiftly.

It is only possible for a human mother to look after more than one needy child because of a culture of food sharing[9] – still almost universally practiced by hunter-gatherers – and other social support. In hunter-gatherer societies, a mother can rely on others to feed her while caring for a newborn or rely on babysitting while she gathers food. Unlike ape mothers, who rarely put down their infants, human mothers do not exclusively look after their own children. Babies born to the

Efé, a nomadic foraging community of central Africa, average 14 different caretakers in their first few days of life[10] – this is 'alloparenting', and
it starts with immediate family such as fathers, older siblings, aunts
and grandparents, and includes in-laws. Human groups are perhaps
unique in recognizing the paternal family, and this key expansion of
our social network is useful for childcare, increases the cultural learning opportunities so that skills and knowledge flow freely and widens
the gene pool for sexual partners. Alloparenting reduces inbreeding
while also providing support and resources for an infant, such as an
apprenticeship during adolescence. It also pays dividends for in-laws
who, though genetically unrelated, have shared interests in the survival
of the next generation.[11]

Cooperation is crucial for the survival of our species. From as early as
three months, infants will selectively choose a helpful over an unhelpful
puppet (during experimental studies),[12] and just a few months later,
they will 'punish' the unhelpful puppet.[13] One explanation for this early
ability to judge character is that human infants – being the only primates that are routinely cared for by different people – need to be able
to distinguish from early on who to trust and who to learn from.

In most hunter-gatherer and pastoralist societies, since the mother
does not have sole care and responsibility for her children, she can contribute to the group's food provision soon after her infant is born. Data
show that women gathering plants, roots and tubers and killing small
animals tend to bring back more reliable calories than men do.[14] And in
many hunter-gatherer communities, including the Nanadukan Agta of
the Philippines and the aboriginal Martu of Western Australia, women
are also hunters. Women continue to play a caretaking role well into old
age. We are the only mammals, apart from orca[15] and short-finned pilot
whales, to experience menopause – most females of other species seldom live beyond childbearing years. This adaptation probably evolved
because of the benefits of the 'grandmother effect', a phenomenon in
hunter-gatherer societies whereby the presence of older women in a
family increases the survival rate of their children and their children's
children. In Hadzabi communities, older women spend more time and
effort foraging for their families than younger women,[16] for instance.

In industrialized societies, too, parents rely on alloparenting, including formal institutions such as schools and, for the birth itself, hospitals

with trained specialists.[17] This outsourcing of care is undergoing dramatic transformation, especially in urban areas, where most of us now live: a local parents' Facebook group lists posts from expectant mothers who already have one or more children, asking for advice on childcare while they are actually giving birth and in immediate recovery. This is a recent phenomenon. For almost our entire history, pregnant women would have had their extended family and friends nearby to help during this time.

Expanding and strengthening our social bonds beyond the immediate mother-child and pair-bond relationships to wider kin and community was an important cultural step for our species, and its origins are likely to be in the social dependency of mothers, the pooling of childcare and the maternal pursuit and maintenance of cooperative networks.[18] It is also a direct consequence of our increase in brain size. This enhanced cooperative behaviour would have benefited our ancestors' survival by increasing the group's resilience in times of trouble, such as during droughts. Over a few million years, our ancestors had become smart, sociable, big-brained people, capable of forging strong, supportive coalitions.

Our ancestors were not the only smart, sociable animals exhibiting cultural skills and behaviours, but while human culture developed, changing our environment and anatomy and driving the further growth of our brains, the others' didn't. Their brains and their cultures have remained roughly unchanged for millions of years. Why didn't the other apes grow bigger brains?

For me, the most compelling answer is that they simply couldn't afford it. Brains are exceptionally expensive to fuel. Neurons need to be kept in a state of readiness, which includes maintaining electrical charge across the membranes, clearing our neural debris and producing new neurotransmitters. Gramme for gramme, brain cells use more energy than body cells, and bigger brains consume more fuel. The human brain weighs just 2 per cent of our body but uses more than 20 per cent of its energy. Great apes can't afford to fuel more neurons than they already have, because they would need to spend an implausible amount of time on foraging and feeding. One study,[19] which looked at the weights, diets and foraging habits of 17 different primate species and calculated the number of neurons for each, concluded that to afford a brain comparative in size to ours, a chimp would have to spend seven hours a day eating and restrict its body-weight to just 26 kilos. At its actual weight, and feeding for seven hours, a chimp could afford a maximum of 32 billion neurons (compared with our 100 billion[20]).

As human culture developed, the cognitive demands increased, spurring a range of evolutionary adaptations to improve energy efficiency and ensure the critical neurons received enough fuel. These included new genes[21] to regulate glucose and creatine transporters (creatine is a backup source of quick-burn energy for when glucose runs low) in the brain, whereas those expressed in our muscles remain unchanged from the primate versions. Evolution optimized our brainpower over our muscles.

Even with these performance-enhancing adaptations, brainpower was limited by the available calories. Our ancestors living during an ice age would have needed to consume at least 3,500 kcal per day just to maintain body heat; other estimates, this time for (slightly bigger) Neanderthals, are 3,360 to 4,480 kcal per day to keep warm and support winter foraging.[22] Paleontologists believe the distinctive broad-nosed face shape of Neanderthals evolved for 'turbo breathing' – to increase their breathing volume and efficiency – which indicates a particularly high-energy lifestyle that would have demanded a high-calorie diet. But apart from honey, fruit or occasional fatty meat, the primate's diet is not routinely high in calories. This is why primates spend significant time feeding and why their brains – and cultures – haven't advanced.

The earliest hominids, such as Lucy, had brains with up to 40 billion neurons, and could have maintained those by eating an ape-like diet continually for seven hours a day. *Homo erectus* (62 billion neurons) would have needed to feed for more than eight hours a day. And later humans, like Neanderthals and ourselves, would have needed at least nine hours a day. That leaves little time for foraging, hunting, socializing and all our other cultural pursuits. Clearly, this was impossible: we wouldn't have had time to fill our bowls with nine hours' worth of food each day, never mind actually eat it. What made our life possible was our pact with fire.

*

One way of understanding life is simply as a chemical system that is able to extract energy out of its environment and power itself. All life revolves around this central energetic relationship. Indeed, natural selection could be thought of as a force that improves energy flows throughout the living world, just as rivers flow toward the lowest point. It is their energy costs that have limited all other animals (and plants) to unchanging roles. A cheetah can run as fast as 120 kilometres per hour (in a short burst) but no faster because of the energy costs to its muscles. Humans, by contrast, have travelled at a top speed of, so far, 40,000 kilometres per hour, in the Apollo 10 spacecraft. (NASA's Juno spacecraft, the fastest man-made object ever recorded, travels at roughly 365,000 km/h.) Chimps are not as clever as us because of the energy costs of a bigger brain. We outsourced those biological costs and in so doing, increased our brainpower.

Let's briefly return to the Big Bang. Everything in the universe has been expanding since then into chaotic disorder – to create order from this, as all life does, requires energy. Plants get their energy from the sun, which emits vast quantities daily. But this is a low-density form of energy. It's enough for photosynthesizing plants to break the strong chemical bonds in the air and make new plant tissue – enough to stay alive, grow and reproduce – but little more. Anything bigger than a few cells, relying on photosynthesis, is capable of very little movement under its own steam. Animals that eat plants get a denser form of energy; those that eat other animals have a denser version still.

Fundamentally, our success as a species comes down to our ability to harness energy and outsource our costs better than any other life form. Instead of relying on our bodies to biochemically break down our foods, we used our culture: we physically processed or predigested (by fermenting and pickling) our foods. Most importantly, though, we used fire to cook.

Just as the act of making fire needs an initial burst of energy – a spark – to overcome the strong molecular bonds in oxygen and the fuel, and enable the fragments to recombine and release energy, so it is a similar situation in our bodies. Food gives us energy, but it takes energy to break down the molecules so they can form new bonds to give us the energy and body tissue we need. In general, eating plants demands much more energy than eating meat for the same gains in nutrition and energy. Cows have to spend hours chewing cud, breaking down long cellulosic chains first in their mouths and then in their four stomachs, before it can be stored as fat. Our brains need high-energy, high-protein foods, and meat and fat give us both. There are still energy costs in acquiring it (through scavenging or hunting), and processing it (using tools, hands and teeth to tear it up for eating), and breaking the molecules down (chewing, digesting and metabolizing), but these are still less costly than for leaves.

Cooked food is much easier to digest because the energy of fire does much of the stomach's work. It is around ten times more efficient to eat cooked meat than raw meat and, kilo for kilo, cooked food provides more calories. That's because the body can absorb more of the food once it's cooked – over 40 per cent more of protein in meat, and more than 50 per cent more of carbohydrates in cereals and root vegetables.[23] Cooking also gives us better access to other nutrients in meat, such as iron, zinc and vitamin B12, which are essential for building and maintaining complex brains.

The invention of cooking also changed the types of foods we ate. Other big animals were less likely to compete for difficult-to-digest tubers or grasses, which meant that they were available for us for little cost. We learned how to process them by pounding and threshing grass seeds to release the edible proteins or grains, and cooked the hard starchy root vegetables until they became calorie-rich, easily digestible foods. And instead of holding large portions of dead animals in our stomachs for hours as, say, lions do, we could digest our food quickly because the fire became a kind of external stomach. With fire doing so much of our digestion for us, over generations our guts shrank, so we can no longer digest many of the raw leaves and fruits that other primates feed on. It was an evolutionary gamble to reduce our food options like this, leaving us vulnerable to famine and unable to tolerate plant toxins other primates can cope with. However, by culling our colons, we redirected precious calories to our bigger brains.

Considering that today hunter-gatherers get a bit over half their calories from animal products and the rest from gathered vegetables,[24] cooking would have made a huge difference to reducing the amount of time our ancestors spent gathering, preparing and chewing food.

While chimpanzees spend around five hours a day chewing, we spend around an hour, giving our species time that other animals don't have. It is easier on the jaw to eat processed food, whether it's been physically, chemically or thermally broken down, and we don't hunt using our bite. This means that there is no selection pressure to retain a carnivorous jaw, and so our mouths, lips, teeth and gapes shrank to – proportionally – around the size of a squirrel monkey's. Through our cultural adaptation to cooking, our jaws became weaker and less prominent, with short muscles that only reach just below our ears (other primates' stretch right up to the tops of their heads), which enables better vocal skills. (This last is important socially, even at the expense of chewing, making the adaptation more likely to spread through the population.) By the time of *Homo erectus* our ancestors had evolved the reduced jaw, teeth and mouth size that made ripping apart raw meat much harder. *H. erectus* had an enlarged, hungry brain that needed high-quality cooked food, and it was smart enough to make it.

So cooking culture was a major driver behind the biological evolution of the human brain: the denser energy in the food that our ancestors ate allowed their brains to expand beyond natural limits and allowed the gut to shrink. These evolutionary changes could have been apparent extremely quickly, because dietary changes can have such an important impact on survival rates. A recent study[25] that looked at Darwin's finches found that after a single drought had limited their available food to just a few hard seeds, those birds with unusually hard beaks preferentially survived to pass on their genes. Of the next generation, only 15 per cent of the group had normal beaks. The change occurred within one year and its effects were seen for 15.

The discovery of cooking would be similarly life changing, and species changing, in a time of very low population, in which genetic differences could have more extreme effects, a phenomenon known as genetic drift. Average life expectancy would have been pretty low – in chimpanzees, it's around 30 years. In times of hardship, such as famine, populations could plummet, threatening the survival of the whole group. Females who were unable to secure enough calories would stop menstruating and become infertile; their babies would be stillborn or die in infancy because their mother was unable to produce breast milk. Those females who managed to get nourishment in difficult circumstances would pass on their genes at a disproportionate rate. Cooking food makes it softer and easier to digest, breaks down toxins and kills bacteria and parasites, making it far safer and more nourishing for weaning infants and children. Cooking would have significantly increased a child's chances of surviving into adulthood.

We know that around 2 million to 1.75 million years ago,* there were enormous environmental pressures because of rapid and extreme climate change, which would have exaggerated the survival effects of small genetic changes, making some traits more likely to persist. Populations may have fragmented, allowing genetic novelties to arise and spread selectively when populations came back into contact. This would have increased diversity. In other words, evolution and speciation would have sped up. And indeed, we see evidence of this in many mammals, including bovids.[26] For our human ancestors, though, harnessing fire and, through cooking, being able to double our calorie intake while reducing our energy losses (fire reduced heat loss at night and passively defended us from predators) was utterly transformative. We became not just novel primates, but an entirely different being: something that didn't just adapt to its environment, but which deliberately adapted its environment to suit itself.

With a new low-cost way of getting calorific glucose, brain size was no longer limited by the ape diet and it rapidly expanded.[27] By 200,000 years ago our brains had reached the maximum possible size defined by our pelvic limitations, although the wiring of our brains continued to evolve efficiencies. However, in recent decades, the advent of safe caesarian sections is having an evolutionary effect. Women whose pelvises are too narrow for vaginal birth are now surviving to pass on their genes, whereas formerly mother and baby would have died. As a result, narrow pelvises are becoming more prevalent: the number of women having C-sections because their pelvises are too narrow has gone from 3 per cent to 3.6 per cent in the past 60 years, a significant increase of 20 per cent.[28] We may one day be as dependent on surgery for birth as we are on assistance. On the

---

* Around the time *H. erectus* emerged and the first evidence of fire making.

other hand, over the past 10,000 years, our brains have decreased in size by around 10 per cent, or by 3 to 4 per cent relative to the size of our body. One theory is that our societies are now so complex that they 'carry' less intelligent individuals who would not have survived smaller groups. However, reduced brain size is common in domesticated animals, and so may be part of the suite of genetic changes connected to our hypersociability and cooperation. It's worth noting that more intelligent people tend to have fewer children;[29] perhaps intelligence is being diluted in the gene pool. Either way, as we increasingly offload our accumulated knowledge into our external brains of stored literature and devices, we perhaps don't need such smart brains to survive.

Nothing brings home quite how reliant the human body is on cooking than the recent fashion for raw food diets. Proponents claim it is healthier because it is how our (distant) ancestors ate. But, despite eating processed foods with far higher calorific content than would have been available even a few hundred years ago, everyone studied on such diets loses weight rapidly and soon switches back to cooked food. Raw fads are not new – the Romans toyed with a kind of Russian-doll feast of raw meat involving a mouse placed inside a chicken, placed inside a peacock, inside a boar, and so on. The diner would sit in a scalding hot bath in order to cook his meal from the outside. Unsurprisingly, this resulted in severe illness and several fatalities, and was mocked by public intellectuals from Juvenal[30] to Pliny.[31]

Indeed, we have become so proficient at processing our food externally, that we can get all the energy and nutrients we require in a concentrated form even without eating animal products. But while we could easily forgo meat, we would struggle to survive if we had to acquire all our food and the fuel to cook it as individuals – and certainly as a 7.5 billion population. While other animals spend most of their waking hours feeding themselves, fire freed us from this yoke, giving our species time to develop our culture, but harnessing us to our group ever more strongly in joint social-cultural enterprise.

And this can turn against our biology. The most recent evolution in cooking culture is proving biologically transformative to entire populations. In the 1960s, the arrival of the TV dinner and other innovations reduced the time taken for daily food preparation and related chores from an average of four hours to 45 minutes. The industrialization of meals has dramatically changed our relationship to food, its origins, and its flavours. Instead of handling the raw ingredients, we simply microwave ready-meals in seconds. Filled with cheap enhancers such as sugar, salt and fats, most of these meals may be disastrous for our health in the long term. Indeed, it is difficult to find foods that have not been sweetened or made salty, with the consequence that from childhood our palates culturally adapt to find unenhanced foods lacking in flavour. Our ancestors would have come across sweet foods, such as honey and dates, rarely, and our biology reflects that the greater risk was starvation, not obesity.

It is extraordinary that in our species, the key survival activity of feeding ourselves relies, like birth, on the help of others, because cooking is a cultural skill that must be learned. Yet it works for us. Now, after tens of thousands of years of cooking evolution, our species enjoys by far the greatest diversity of meals and our genes have adapted in turn. People whose ancestors farmed have different concentrations of saliva enzymes and gut bacteria – ones better suited to breaking down starch – than descendants of non-grain-eating hunter-gatherers. Hunter-gatherers' guts are exquisitely tuned to their environments with a microbiome that swings through annual cycles. Similarly, populations with a cultural history of consuming milk or alcohol have genes that enable their bearers to better digest them.

5

Cultural Levers

In 1860, a grand expedition of 19 men, 26 camels, 23 horses, and six wagons left Melbourne on the south coast of Australia to discover the best route for an overland telegraph line through the uncharted interior of the continent to the Gulf of Carpentaria in the north, a distance of around 3,250 kilometres. Robert Burke (an ex-army officer and police superintendent) and William John Wills (a surveyor) led the expedition, which set off with great fanfare from the Royal Park, watched by more than 15,000 spectators.

There were early signs this wasn't going to be the greatest success. Their six wagons, loaded with food to last two years, assorted furniture, and, inexplicably, a Chinese gong, weighed 20 tons, and one of the wagons collapsed before they had even made it out of the park. It took them three days to reach the outskirts of the city, by which time another two wagons had broken down. By the time the group reached Cooper's Creek, the furthest limit explored by Europeans at the time, they had ditched the majority of their load, including the 60 gallons of rum, purportedly for the camels to prevent them getting scurvy. Here, the group split and just Burke, Wills, Charles Gray (a sailor) and John King (a soldier) set off for the northern coast with three months' supply of food in the height of summer.

Burke was highly suspicious of the indigenous people they met, shooting over the heads of natives that offered them fish to eat, and he ordered his expedition to keep Aboriginals away. Fifty-nine days later, weak, short of food, and with swamps blocking their way, the group decided to turn back. Before long, they were eating their pack animals. Gray soon died of dysentery while the remaining trio eventually made it back to Cooper's Creek, expecting to rejoin the rest of their party, only to find the camp had been abandoned hours earlier.

Disaster continued to befall the trio until they met a local Aboriginal tribe, the Yandruwandha, who looked after them with fish, beans, and their staple bread, made from a kind of seed called ngardu. However, Burke remained antagonistic toward the Yandruwandha and eventually drove them away by shooting at one of them. As the hapless explorers continued their journey, they sought out more of the ngardu seeds from a semi-aquatic fern. Initially, the men tried boiling the seeds, until they came across some Aboriginal grinding stones, which they used to grind a flour. Delighted, the trio spent a month eating five to six pounds of ngardu bread a day, but mysteriously they grew weaker and weaker, suffering painful bowel movements that 'seem[ed] greatly to exceed the quantity of bread consumed, and is very slightly altered in appearance from what it was when eaten'. Within a week of this diary entry, both Wills and Burke were dead. The remaining member, King, managed to survive by appealing to the Yandruwandha, who took him in. Three months later – time enough for him to have conceived a child with one of the Yandruwandha women – a relief party from Melbourne found King and brought him back to the city.

Like so many European explorers, Burke and Wills fell foul of the cultural knowledge trap. If they had tapped into the accumulated wisdom of the people indigenous to the area, they would have learned how to prepare the ngardu to be nourishing rather than deadly. Ngardu must be collected once it has aged, not while it is young and green, and it must be ground for digestibility and then sluiced thoroughly with water. This leaches the flour of thiaminase, an enzyme that, if ingested, destroys the body's vitamin B1. The trio would have learned to expose the ngardu flour directly to the ash during baking, which further breaks down the enzyme. Without such cultural knowledge, the explorers unwittingly poisoned themselves.

It is so easy to look at the simple necessities in our lives – our foods, our clothes, our basic tools – and believe that, if push came to shove, we could make them ourselves. We are, after all, the most intelligent animals on Earth. Yet, it isn't our individual smarts that have got us here.

We've seen how outsourcing our energy requirements allowed us to change our environment and bodies, and build our brains. Now, let's look at the cultural levers that enable us to outsource the energy costs of our activities: we extend our physical capabilities with tools, and outsource the cognitive costs of problem solving to the collective brain of our group. Through cumulative cultural evolution our species increases its population, with successful exploitation of the environment, in the most energy-efficient way – it is entirely under-pinned by cultural levers.

Technology has allowed our species to be a highly efficient planetary operator, capable of deploying vast amounts of energy while expending no more muscle power than the tap of a finger. What then are we using . . . our minds? Yes, and no. The physical lever that enables a puny primate to level a mountaintop is coupled to a cognitive lever: humans rely on their society's collective brain for their tools, behaviours and skills, from fire making to cooking – we rely on it just to survive.

Local knowledge is indispensable because of an evolutionary trade-off, in which our species gave up innate adaptation to an ancestral habitat in return for the culturally adaptive versatility to survive any environment. Liberation from our ecological niche came with a cost to our self-reliance: we cannot biologically adapt to every environ-ment, so we have to rely on others for survival knowledge.

Cultural knowledge accumulated over generations allows com-munities to glean information, read landscapes and find food and shelter easily. The Yandruwandha were able to see the food around them that was invisible to the Europeans, just as a European dropped into a city would easily locate cafés. We are all experts at navigating our own environments – we've been learning since infancy. Your cultural developing bath – the behaviours, technologies and other cultural practices of your society – shapes your behaviour, cognition, percep-tion, personality, intelligence, physical abilities and more, just as the chemical composition of a photographic developing bath generates a unique picture.

We can see evidence for this in our neurology – our brains are liter-ally shaped by our culture. One recent study looked at the intelligence-governing folds in the cerebral cortex of hundreds of human and chimp brains. These wrinkles, called sulci, continue to grow and change after birth, but differently in the two species.[1] In chimps, researchers found, the shape and location of these brain folds is largely determined by genes (siblings had near-identical folds), whereas in humans, genes play a far lesser role, allowing environmental and social factors to shape our brains. Since chimps are to a greater extent genetically locked into their cognition, they are limited in how their brains can develop and their capacity to learn new behaviours or skills, compared with humans. Humans are anyway born with less developed brains than chimps, allowing more of their brain growth to occur postnatally, where the outside world plays a larger role.

This extraordinary plasticity of mind propelled our ancestors' intelligence and cultural development; however, it means we need to learn from others almost everything to survive. The demands of cultural learning require an exceptionally large brain, a long childhood and adolescence in which to learn, and a strong social group – characteristics that evolved in symphony as a successful strategy. Our first teacher is our mother, for whom we have an innate attraction, recognizing and seeking out her voice and her face, and following her gaze automatically from birth. As we grow older, other family members, peers and older, trusted group members become our teachers.

We are now so programmed to use social resources to solve life's problems that we seldom attempt to work things out for ourselves and are quick to ask others for help, whereas chimps will not. Solving a problem by copying a solution from others usually requires less physical and mental labour than using trial and error. While chimps must individually work everything out for themselves – reinvent the wheel each time, so to speak – we can rely on the efficiency of our cultural evolutionary process to have produced the best way of doing things. Not only are chimp brains smaller and less powerful, they also have to do more work to solve the same problems, which gives them less cognitive capacity to combine skills and produce more complex culture.

Of course, we can only rely on the collective knowledge of the group to hold the solution to our problems because of the successful copying mechanism that cultural evolution itself relies on. Copying is the basis of cultural evolution, just as the copying of genetic sequences is for biological evolution. If we do not copy precisely enough – with enough fidelity – then different cultural practices will not persist long enough in the community to be replicated, and cumulative culture cannot occur.[2] High-fidelity cultural transmission greatly increases the longevity of cultural variants within a group and leads to a population with a far richer, more diverse culture.[3] This is because the more accurately something is copied, the more versions of this practice exist in the group, and the more opportunities there are for tiny modifications and refinements of these practices to evolve – the mutations that lead to variations.

By copying, we made our world. It may seem incredible that there has been no designer behind our cultural solutions, the practices and technologies we use. We are used to associating inventions with inventors – Thomas Edison famously invented the light bulb and Johannes Gutenberg the printing press. But in reality nothing was truly invented by a lone genius. Innovation and discovery are usually the result of chance or iterative refinements and combinations of existing technologies – the Darwinian process of blind variation and selective retention.[4] Indeed, in models of how cumulative culture builds in complexity, the rate of invention (of new traits) has the least effect on the number of innovations, whereas the rate of *combination* of existing traits has the greatest.[5] Accurate copying ensures that a practice can circulate in a population long enough to be combined with others, allowing cultures to evolve in complexity and diversify under a process of natural selection.

Nevertheless, it seems counterintuitive that we would evolve such large brains only to use them primarily to copy each other, and for many experts, the question persisted over whether invention or copying is the best for problem solving. After all, by engaging directly with our changing environment, like primates often do when they are trying to solve a problem, we would be acquiring firsthand, up-to-date and relevant knowledge.

In 2010, evolutionary biologist Kevin Laland set out to answer the question experimentally. His team designed a computer tournament in which avatars had to navigate and survive a strange world – somewhere between *Survivor* and *Second Life* – to win a £10,000 prize. More than 100 teams competed, including neuroscientists, computational biologists and evolutionary psychologists. They programmed their avatars to survive the unfamiliar, changing environment. Laland, like most of the experts in his field, expected that the best survival strategy would involve a mixture of innovation and copying.

What they found astounded them: copying beat innovation hands down for all plausible conditions.[6] 'There's no balance, no mix of asocial [innovation] and social learning,' Laland said. The tournament was won by two graduate students, a mathematician and a neuroscientist, whose program used a copying strategy that preferentially copied more recent actions when the environment changed faster, meaning their avatars paid less attention to actions that may have become outdated. We humans are also strategic, selective copiers. We choose whom to learn from in different circumstances, keeping ourselves up to date and reliably informed.

No individual had the grand idea to perform a seven-step processing routine for ngardu. Instead, the method evolved iteratively over generations, each improvement copied more frequently by others until the best way of making bread emerged as a ready-made cultural practice to learn. And yet, even when a cultural practice proves successful enough to survive the generations, its transmission to others in the cultural developing bath may be under the auspices of tradition, which is also cultural, and may not on the surface have anything to do with a practice's actual advantage. The Yandruwandha grind and sluice the ngardu for ritualistic reasons, even though its laborious preparation (*pita-ru,* meaning 'always pounding') is the lament of hard-working women. It is only recently that scientists have discovered how the steps selected by cultural evolution greatly reduce the risk of thiaminase poisoning.

We do not have to understand *why* each step of a practice is important; we need only to learn the steps, a key difference between humans and other intelligent animals. There's a very telling experiment, done by evolutionary psychologist Mike Tomasello at the Max Planck Institute in Germany, in which a puzzle box containing a treat is given to a human toddler and a chimpanzee. Neither is able to get the treat out. He then demonstrates a multi-step process of pulling and pushing pegs that eventually releases the treat. Among the motions, he includes an obviously nonsensical step – patting his head three times before the last step. Both toddler and chimp are able to copy his actions and get the treat, but only the toddler includes the head-patting step. The chimp, seeing this is not relevant to getting the treat, omits it from the routine. The human, however, unquestioningly copies all the steps. The toddler trusts the human teaching her to have a reason for each step in this situation, and so she overcopies. In fact, the less clear the goal of the procedure is, the more carefully and precisely the human child will imitate even irrelevant steps.[7]

Copying is so essential that we have evolved cultural and biological mechanisms to favour it, including longer childhoods, wider social groups and better memories. We also teach. A human mother will show her infant how to master a task, and she will watch as he copies her. She will modify her demonstration to his attempts and help him achieve the task in a two-way responsive process, before showing him the next step. Other animals do not do this.

Teaching allows knowledge to be transmitted in a very accurate way and the student learns far more efficiently than by copying alone, especially for skills that are complex or require a precise sequence of actions. In one study[8] comparing methods of learning a stone-tool knapping technique, teaching doubled the performance of other methods of cultural transmission. Perhaps it was teaching that gave humans the sufficiently high-fidelity transmission mechanism to enable cumulative culture. The stone-knapping study points to a reason why our early hominid ancestors were stuck in a technological stasis for more than 700,000 years, making the same primitive Oldowan stone tools. More complex Acheulean stone tools required a longer sequence of knapping steps, and copying alone wasn't a sufficiently reliable method of social learning. People needed to be actively taught how to make Acheulean tools, and so they didn't appear until our brains could handle it – with *Homo erectus,* around 1.8 million years ago.

But teaching is costly for the teacher, and it only evolves where the benefits of acquiring valuable information outweigh those energy costs. For smart animals, like chimps, it is not worth adults investing in the costs of teaching because youngsters are clever enough to pick up the relatively simple skills they need. Teaching is an altruistic act that occurs in a few species that also practise alloparenting (cooperative breeding), such as ants and meerkats. Cultural complexity relies on teaching for sufficiently accurate knowledge transfer, but it also makes teaching a more cost-effective method of cultural transmission. That's because as cultural practices become more complex, knowledge becomes more valuable, and relying on copying alone becomes inefficient and unreliable. Also, the pool of teachers increases because cultural knowledge accumulates in the population as it grows in complexity, so more people then possess enough knowledge to pass onto students. Teaching explains cultural complexity but is also the product of it, in yet another human evolutionary feedback mechanism.

The practices and technologies that accumulate in our cultural toolbox emerge through countless iterations as they are copied over generations. Environmental change can trigger a burst of cultural variation, just as it does for biological evolution. For instance, researchers have connected a series of major climatic and landscape changes that occurred in East Africa around 320,000 years ago with the emergence of complex cultural traits such as the manufacture of and trade in sharp obsidian blades.[9] Necessity is not necessarily the mother of invention, but this serves to demonstrate how new selection pressures acting on existing technologies and behaviours can alter their rate of transmission. If terrestrial prey became scarce, a formerly rare skill in fishhook-making may become more prevalent,[10] or, as occurred during an expansion of grasslands 65,000[11] years ago in Australia, seed grinding. It's sometimes helpful to think of evolution as the failure of the least fit, rather than survival of the fittest. Among the diversity of processes and techniques, some will fail, becoming rare or eliminated by society over generations. The rest will continue to be copied and available to the social group, giving populations adaptive flexibility.

Environmental change also affects population size, which has its own important effects on culture, because it changes the size of the collective brain. Just as the collective brain acts like a lever to make learning less effortful for the individual, so the longer that cultural lever is – the more cultural practices it includes – the more energy-efficient *the group* becomes, and so cumulative cultural evolution accelerates. Because innovations often arise from combining existing ideas, just a few more versions in the collective toolbox can have an enormous impact, because so many more combinations become possible. Consider that three items can be combined in six different ways (if each is only used once), whereas four can be combined in 24 different ways, and for ten there are more than 3.5 million combinations. Only larger groups have the collective brain capacity for a big toolbox and equally, only larger groups can afford the physical energy costs to benefit. As a result, there comes a tipping point as populations scale up when greater complexity suddenly takes off, and we see cultural explosions.

One such creative explosion seems to have occurred around 40,000 years ago in Europe, judging by innovations in the archaeological finds. It is this that led some experts[12] to claim that modern human culture, including complex language and tools, emerged then. The suggestion is that a genetic modification around this time – perhaps resulting from interbreeding with Neanderthals – produced a cognitive boost in our archaic ancestors, generating a breed of behaviourally modern humans. There is no compelling evidence to support this. The reason we see a wealth of artifacts at that time in Europe is not because these people were any different. In part, it is because those sites have been studied more extensively over the past centuries, and they are cooler, drier, protected environments – often caves – that preserve ancient materials better than in the tropics.

The other reason, though, is that demographic, social, environmental and cultural changes occurring in Europe at this time were driving cultural complexity. The greatest population boom in prehistory occurred 40,000 to 50,000 years ago, geneticists recently discovered.[13] Meanwhile, another team of geneticists, who were comparing the cultural explosions that happened in Europe 45,000 years ago and in sub-Saharan Africa 90,000 years ago, found great similarities between their respective population densities.[14] Larger, more culturally diverse populations can call upon a greater resource of potential solutions as the physical or social environment changes. This gives them more opportunities to adapt their cultural practices, making their societies more resilient, so their tool and artifact methods have a better chance of surviving and increasing in complexity. The bigger the population is, the longer the cultural levers.[15]

Similarly, the better a group is connected to other groups – and the better the connections are within a group – the more chance individuals have of acquiring new cultural practices and technologies. The reverse is also true: small, isolated communities can experience a cultural evolution towards simpler, less diverse technologies, effectively losing culture. Sometimes even a fundamental technology can be lost.* It follows, of course, that any cultural practice that increases a society's population – improving nutrition, fertility, or reducing infant mortality – will produce more carriers of that practice, so they will spread faster and further. In this way, technologies like fire making rapidly become universal.

From the loss and gain of cultural technologies, whole societies propagate that rely on the structures of the group connections that enabled those technologies. I am currently typing this on a computer. I don't need to know how each key was fashioned out of plastic, painted with a letter, and fitted into the keyboard. I don't need to know anything about how the letters appear on the screen, only that I must tap the keys with my fingers for the letters to appear. I am relying on a complicated network of thousands of individuals, including engineers, craftspeople, factory workers, miners and more, without whom this would be impossible. That is what is means to live in a materially and culturally complex global society; I couldn't possibly know all the steps to get here, let alone re-create them in my everyday life. What is more remarkable is that none of the people involved in this myriad of activity does either. A miner learns the correct striking angle of his chisel and which rocks to hit – he does not know if the lumps he extracts from the earth will go on to be processed for a ship's hull or an electronic component. Just as we see the enormous variety of ecology and complexity of life that has emerged from biological evolution, so has cultural evolution built entire systems upon which our everyday practices continue to play out.

I hold in the palm of my hand a lump of rock – flint, formed from the death of microscopic sea creatures that, during their lives, transformed the energy from their food into the making of skeletons that

* I explore this in detail in Chapter 10.

were themselves transformed into a type of quartz that would, millions of years later, be spat into cliffs by the tremendous energy of tectonic movement. This teardrop of flint has been further transformed, this time by human hands: I hold in the palm of my hand an axe, a tool made more than 40,000 years ago. From the stuff of our physical environment, a living creature crafted something original, inert, yet capable of doing work. My hand, as human as the maker's, is of a similar size – the axe fits perfectly in my palm. I feel its weight and the deliberate ergonomic shaping of the tool, instinctively folding my fingers around its indentations. Had I been taught how, I could use it to scrape meat from a freshly killed deer, which is perhaps what its last job was.

The hand axe was the Swiss Army knife of its time, an indispensable and versatile tool chipped from stone and used for chopping, slicing, boring, shaping, bludgeoning, chipping, carving wooden tools and so many tasks that would otherwise have taken much more labour. It is, in other words, a physical cultural lever.

The earliest handaxes we've found date back over 1.5 million years, and they have been used more or less continuously up until the last century in some hunter-gatherer communities. We find them from sub-Saharan Africa to the Arctic, left in caves or amassed in factory-scale production facilities at the bases of cliffs. Given how key the handaxe was to human survival, it is surprisingly difficult to make, and individual crafters rely on others to teach them the stone-finding, extracting and crafting skills. Bear in mind that the average human toolbox at this time would have included a variety of specialist stone and wooden tools, twines and other hafting materials, flint fire starters and tinder, as well as skins, gut and other animal products. The term 'Stone Age' is often used to mean primitive or backward, but even hundreds of thousands of years ago – and certainly by the time of our own species – stone working had developed into a highly sophisticated technology requiring skilled craftsmanship and a knowledge of geology, fracture mechanics and the thermal properties of rocks. Anthropologists recently discovered sophisticated stone spear tips in South Africa, made 500,000 years ago[16] by *Homo heidelbergensis*. Compound tools like these required knowledge of different materials that had to be fashioned separately, including the wooden shaft and twine to fasten the stone tip. To attach the spear securely, resin glue (from the bark of certain trees) would have had to be softened in a fire.

Making compound tools is cognitively demanding – too demanding for other animals – because it uses the brain's 'working memory' to retrieve, process and hold in mind several chunks of information at the same time. Working memory is used for multi-tasking and strategizing, and a variety of early technologies required this sort of brainpower, including animal snares and traps. Making a trap means imagining and creating a device that can snag and hold an animal, and then returning later to see whether it worked. There are also physical demands: concentrating for the period required is tiring, and gathering materials and sculpting a tool is exhausting – and that's if you've been taught how to do it and are competent. Innovation relies on trial and error – many hours of wasted energy expended for, hopefully, the same eventual result. Animals like us get their energy from metabolizing food, so greater energy expenditure requires more food, which itself takes time and energy to acquire. Once you've been shown how to do something, and once you've practised a technique enough for it to become second nature, the energy required falls considerably.

The efficiencies of time and energy that come from such reliable (high-fidelity) copying allow our technologies to accelerate in complexity, generating more energy-saving devices and specialist equipment (if you've ever used a knife to tighten a screw, you'll understand the efficiency benefits of the right tool for the right job). This required a significant investment in time and energy for individuals but also for the group as a whole. The energy outlay only makes sense with the efficiencies of scale that come with larger groups who can spare the specialist labour (physical lever), and indeed, it would only have been possible for larger groups who held the collective knowledge (cognitive lever). Some of this scaling effect can be achieved through the establishment of good dependable networks with other groups, before the groups themselves become bigger. Collective knowledge can be pooled in this way, and trade in resources and skills enables each group's labour costs to be reduced. This is why larger and better-connected populations produce fancier technologies.

The cultural evolution that drove our technological complexity relied on our ability to outsource much of the cognitive processing, memory, knowledge, and physical labour to the group, resulting in a production capacity that far outstrips our biological capabilities. Energy efficiency was a strong selection pressure acting on our cultural evolution as well as our biological evolution. Gradually we decoupled our individual physical and biological capabilities from the human ability to transform the environment. The invention of weapons and food-processing tools meant we could do away with the big jaws, teeth and claws of other carnivores; our social tools, though, would allow us to do things far beyond the biological capabilities of any other animal. Everything we have now, from fire to the paperclip to the iPhone, arose from our species' increasing mastery of efficient energy processing.

As our technologies have evolved, our physical cultural lever has lengthened: our daily 2,000 food calories give our bodies enough energy to power 90 watts (at an average metabolic rate), and yet the power we wield goes far beyond this. To put this in perspective, 90 watts is enough to power an incandescent light bulb. As I write, I'm sitting underneath two such bulbs, with a desk lamp behind me, in front of a working computer and monitor. The radio is on, as well as a fan heater. The washing machine has just gone onto spin cycle, and most of my day's calories are being cooked in the oven. This is clearly not the work of my morning porridge. The average British person now uses four times our metabolic power just in home power; a US national uses nearly 12. Humanity in total now uses around 17.5 terawatts of power, which would make us responsible for 2,300 watts each – 26 times our 'natural' ability. We have achieved this phenomenal leverage by outsourcing and distributing energetic and time-consuming activities. This allows us to generate a surplus of energy, food and time, resulting in greater populations, which enables further efficiencies in energy and resource use through the economy of scale. Distributing labour, for example, gives specialists more time and energy to accelerate the evolution of cultural practices. Our physical cultural levers evolved in efficiency and scale until we reached another tipping point, in which labour-intensive processes from food acquisition to transport became so cheaply and freely available that we were able to become planetary operators. Humans now use more than

40 per cent of Earth's total primary production, which is all the energy captured from the sun by plant life (and thus available for the rest of the planet's life).

A key driver of cultural evolution is that a new practice will improve the production or flow of energy, and thus improve the survival of our genes – having offspring is energetically costly and, for every animal, their metabolism limits their fertility.* However, our cultural evolutionary success has ultimately decoupled our cultural survival from the survival of our genes. It is a curious fact that the wealthier an industrial society becomes, the fewer children its people produce – some fertility rates are so low that population rates are actually declining despite the best availability of food and health care. We are, through our cultural evolution, overruling the key testament of biological evolution.

Just as we harnessed energy to change our landscapes and ourselves, so we use it to change the stuff of the natural world into the stuff of our human world – almost everything we use and are surrounded by in our daily lives is made by us, and we rely on this artificial infrastructure to manage the energetic and social flows of our lives. When we describe something as artificial, it is something taken from nature and rearranged by us, and what are we if not part of nature? Our cultural evolution is part of our biology just as its products are a part of the new Earth we've helped create.

Birds make nests and beavers make dams – rearranging the stuff of our environment into useful technologies – but only humans have taken the raw constituents of the world and from them guided the material evolution of diverse and complex products. Because our technologies evolve through combination, and because our societies and cultures are so interconnected, one discovery or practice can propagate countless others, across our web of social and technological reliance, and we have the intelligence and plasticity of mind to recognize and respond to discoveries. Take mud. From mud, you can make almost anything, and people have. Fire makes it permanent,

---

* One theory for why our own babies are born so immature (compared to other primates) is that the metabolic costs for the mother of fuelling herself and a bigger foetus become unsustainable around 40 weeks of gestation.

turning the pliable molecular sheets into a solid three-dimensional object with completely different characteristics. Firing clay was as transformative for our culture as it was for the material.

Pottery provided a way of cooking stews and broths; preparing fats, seafood and fermented beverages; and carrying liquids. Before pottery, the only way nomadic peoples could carry or store water was in bladders or skins, so a solid container for blood, milk, water, oils and entrails would have been revolutionary. The ability to make soups would have helped wean infants, allowing them to eat a wide range of nutritious, easily digested, detoxified foods, providing gradual exposure to new and potentially dangerous foodstuffs. Fish broths prepared in pots, for example, retain the fats, including omega-3 lipids for infant brain development and female fertility. Soups alone would increase childhood health and survival rate, resulting in real impact on population numbers.

Pottery may have been the technology that made agriculture possible – it is hard to imagine how grains would have been stored, cooked or fermented before. There is a clear explosion in pottery culture contemporaneously with agriculture everywhere.[17] Storage also had a lasting impact on the social structure, territoriality and economy of egalitarian[18] hunter-gatherer societies, because stored foods can be owned and redistributed, presenting opportunities for political manipulation.

Pottery was the first material that humans changed from a natural material to an artificial one, and it reveals the feedback relationships that unfold between societies and their inventions as each change drives further possibilities. Over millennia of cultural evolution, the techniques[19] for making, firing and decorating pottery evolved in complexity and diversity across the world just as the products themselves, including milking jars, figurines, bricks, roof tiles, lamps, toilet bowls, ceramic electronics, and so much more. The most expensive part of pottery is the firing – fuel needs to be gathered, the kiln needs to be kept hot enough – but several pots can be fired at the same time. This mass production made it cheaper, so pottery quickly replaced rival technologies like basketry or wooden-box making, where only one container can be made at a time.

As societies controlled more energy, so their technologies evolved to be able to do more work with greater efficiency. The kiln technology that developed for pottery, to produce the controlled hotter temperatures for glazes, probably conceived metallurgy.[20] Rocky minerals, crushed and used for decoration, would have deposited little beads of copper in the fire bed that could be beaten and melted. The discovery that copper could be obtained from rocks – by smelting mineral ores such as bright green rocks of malachite, covellite and copper sulphide in a hot fire – must have been pretty mind-blowing. Suddenly the inert stuff of the planet, the ground beneath us, was found to be hiding incredible new materials that could be shaped into anything and then reshaped to be used again and again.

For this, people needed yet more energy: temperatures of at least 1,000 degrees Celsius in a kiln fuelled by charcoal and supported with bellows. With hard new copper blades, people could cut bone, wood and even stone – the great pyramids of Egypt were made by slaves carving stone blocks using copper chisels. To create the estimated 300,000 chisels needed, around 10,000 tons of copper ore[21] would have been mined in conditions so appalling that life expectancy for a miner was less than a year.

Adding tin to the copper, people discovered by 3000 BCE, gave them bronze,[22] a harder alloy.[23] Bronze opened new trade routes, because tin is relatively rare in Earth's crust and had to be sourced as far away as Cornwall in England and across to Afghanistan along tin routes that spread ideas as well as commodities. It was the first large-scale international trade network and it made a new class of elite very rich.[24] When these trade routes were abruptly disrupted in 1200 BCE by a wave of invasions by nomadic societies, people looked for an alternative to bronze. They found it all around – pretty much every rock contains some iron, the most democratic metal. We entered the Iron Age and have never left it.

Smelting iron ore requires still more energy – a much hotter fire than for copper. The best people could achieve with ancient furnaces was a holey, spongy mass called bloomery iron, which wasn't much better than copper. Hammering it out improved its strength but it was still a poor substitute for bronze. (Nevertheless, wrought, or beaten, iron was common in ancient Egypt by 1500 BCE.) The breakthrough came when forgers worked out a way of making and controlling hotter fires by fuelling them with charcoal and, in doing so, created an alloy of iron and carbon,[25] which we know as steel. This was by far the strongest metal yet, except when it wasn't: it turns out the amount of carbon in the alloy is critical – 1 per cent carbon makes good strong steel; 4 per cent carbon makes it weak and brittle. But it wouldn't be until the twentieth century that we would understand why some steel-making processes worked and others failed.

Until then, successful methods were passed down through the generations as closely guarded, highly complicated secret rituals. When the Romans left Britain, they carefully hid their iron nails and other metallurgy, protecting the knowledge that enabled them to make swords that wouldn't snap, aqueducts and ships. A pit discovered in Scotland contained a seven-ton hoard of iron and steel nails buried by one such retreating Roman legion. The loss of the key technology of steel making became mythologized in British legends of unbreakable weapons, such as Excalibur, the magical sword of King Arthur.

The blast furnace – a technique of heating the ore in a fire with charcoal to reduce it, and then blasting it with air – was invented in many incarnations in several places around the world and is still used today. It was the manufacture of this extraordinary, ordinary metal that allowed iron tools to create our modern world. Iron-dipped ploughs cultivate more land faster; iron axes chop trees and clear land quicker than stone; iron nails and iron aqueducts and bridges make for stronger infrastructure; all of which led to the rise of towns and cities with more people. However, in our bid to control ever more energy, we change the environment that made us and that supports our societies. The demand for charcoal for metallurgy has caused widespread deforestation and environmental damage across the world, usually with socioeconomic consequences.[26]

There is no way that an individual could chance upon the discovery or, however clever, invent for themselves a way of conjuring steel from rock. The various techniques each require many steps that have been learned and passed down over generations. Such complex culture relies on a society that values teaching and learning, and that has strong networks across geographical areas. It is a society that is big enough to have division of labour, and that can feed and water its labourers sufficiently. Today's world only exists because enough time has passed for technologies and societies to evolve in complexity, and for populations and networks to grow big enough to support the energy costs involved.

Making and controlling fire gave humans an amazing ability to transform the stuff of our planet into the materials of our manmade world. Fire-mastery marks a turning point in the story of us, but also in the story of life on Earth, because it was the first step towards becoming a new planetary force. We changed forever the energy dynamic between a living organism and its environment, and we did it almost entirely through strategic copying of each other, building together a collective smart brain.

WORD

Evolution is wholly reliant on the transfer of information between individuals – information that has been faithfully copied, stored and transmitted. In biological systems, genetic information is encoded in DNA. For human cultural evolution, the essential information – cultural knowledge – is encoded in words. Just as biology has evolved strategies that improve the reproductive process for its genes, so too culture has evolved adaptations that improve its reproduction.

6

Story

In firelight at the edge of the ocean above the wave-slapped sand, a man is singing to me and not to me. He is rising and crouching, to-ing and fro-ing in the flickering light, his black skin disappearing into the night and his body paint gleaming brilliantly, until I am awestruck in the presence of a dancing, writhing spirit creature with flailing arms, rhythmically stamping feet and flashing eyes and teeth. He sings and strikes painted sticks as his feet slam into the red earth, vibrating the warm ground beneath us. A decorated teenager blows music into the buzzing didgeridoo, and the elder dances more wildly, throwing back his head and clawing the air in a chaotic yet contained rhythm. The fire crackles and others around me are join-ing in the song, hitting sticks and shaking dried seedpods. Hours pass and the Yolngu elder is still dancing and singing. He will sing all night long until the morning star rises.

He is singing the story of Creation. It's about the Dreamtime, when the first people were carried to Australia by the creator spirit, Bar-numbirr, on journeys across land and sea. As she flew, Barnumbirr – known to us as Venus, or the morning star – sang a song of her journey, describing the landmarks on her route and telling the story of the beginning, of genesis. The song, the dance, the ceremonial body paint are entrancing, imprinting a vivid, lived impression on my mind. The rhythm of the stamping, the clapsticks, drumming and didgeridoo, the flashing firelight, the repetitive haunting song, the sense of a meaningful, shared experience – it is unforgettable. And indeed, songs like this haven't been forgotten. They have been taught, learned and passed down the generations for thousands of years. Perhaps for 60,000 years, since the first people arrived in Australia. They are the songlines.

Stories like the songlines are oral archives of cultural knowledge that bind their collaborators together through shared cultural reference, subtly redefining the parameters of family or society. Each Aboriginal group has its own unique songlines, detailing their laws, ceremonies, duties and responsibilities, as well as their spiritual ancestors and landscape. Songlines are also living, story-bound maps – they are the invisible pathways that crisscross Australia. Melodic variance, artworks and dance are used to describe landmarks, trees, rocky protrusions, creatures, weather patterns and water holes – often with reference to the constellations above. This allows songlines to transcend the many different language groups. If you know the song, you can find your way from one end of a track to another: each musical phrase is a map reference, 'a memory bank for finding one's way about the world', as Bruce Chatwin wrote in his seminal study.[1]

This explains the prevalence and importance of human stories: they work as collective memory banks, storing detailed cultural information encoded in narrative. Stories help cultural knowledge to linger in the collective memory long enough to accumulate and evolve, and they provide a reliable, energy-efficient[2] way of transmitting complex, context-rich cultural information widely. As human culture evolved in complexity, storytelling became more than a vital cultural adaptation – our brains evolved with reflexive use of narrative as part of our cognition. Stories shaped our minds, our societies and our interaction with the environment. Stories save our lives.

It was a relatively small band of pioneering Australians that landed on the continent some 65,000 years ago,[3] spreading rapidly, establishing thriving clans and communities, learning to make their unique environment work for their survival. They used firestick farming and made complex tools, including fishing harpoons and hunting spears, from multiple materials. Clans moved frequently, following the changing dry and rainy seasons, travelling between water holes and other resources, mapping the land intimately as they went. Stories are a useful technology for learning, recalling and teaching. As one Aboriginal elder explains: 'We have no books, our history is in the land. We learned from our grandmothers and grandfathers as they showed us these sacred sites, told us the stories, sang and danced with us the Tjukurpa (the Dreaming Law). We remember it all in our minds, our bodies and feet as we dance the stories. We continually recreate the Tjukurpa.'[4] The cultural techniques, passed down through the generations in songlines, led to flourishing human populations across Australia.

Storytelling is an inherently social enterprise – it relies on people sharing a mental commons, agreeing together to suspend reality and explore a virtual space-time. Although the songlines allow Aboriginal groups to differentiate from each other, they also, crucially, function as a unifying factor. These extraordinary oral maps of stories, land, people and culture are not only essential to indigenous identity, they probably saved the Aboriginal people from extinction.

Around 20,000 years ago, a trenchant ice age devastated Australia's environment. On the other side of the planet, the Eurasian ice sheet extended 4,500 kilometres across, and on its own lowered global sea levels by 20 metres, locking up so much water that rains failed across the world. As the droughts became more severe, conditions became impossible for many mammals: Australia's giant marsupials all died out during this time, and the human population crashed by 60 per cent. Those groups that managed to cling on were increasingly isolated in geographically distant refugia across the vast continent. And the situation persisted for thousands of years. Small, isolated populations experiencing incredibly challenging environmental conditions deliver the classic ingredients for an extinction: with the gene pool not being sufficiently refreshed, devastating mutations creep in and weaken the population.

What should have been an evolutionary dead end – a human population isolated from the rest of the world's people for tens of thousands of years, then split into tiny, isolated groups – did not result in local extinction.[5] How did Aboriginal Australians survive, when so many other big animals died out?

Songlines saved them. During this period, facing especially harsh environmental challenges, people had to rely to a far greater extent on specialized knowledge to find the resources they needed and navigate different environments. Grinding stones dating to this ice age period reveal that people were already skilled in processing ngardu,[6] and the discovery of adult molars with specific wear patterns point to people processing fibres to make fishing nets. These multi-step, complex techniques had to have been stored in the collective memory bank, and passed on, even when such information was useless – when there was no ngardu growing in the area a group was living, for example – to be recalled, as a lifesaver, perhaps generations later.

And the songlines also helped ensure there was a healthy population to host this cultural information, rather as our 'selfish genes' drive their own propagation. Throughout the terrible ice age, songlines and the rituals they describe helped tribes cope with isolation, and isolation helped the stories and rituals survive. Without an influx of different people with different ideas, there's less pressure for a culture to change. However, because songlines could be universally understood, they also allowed for some intergroup connectedness – songlines operated as mating networks, allowing for the necessary genetic exchange that ensures diversity and staves off extinction. Songlines kept the cultural and gene pools healthy, enabling ice age Aboriginal culture to find a balance between being separated and connected, which had eluded the other large mammals. As the climate warmed and the continent became more habitable, Aboriginal populations flourished. There were around a million people living in 300 different language groups by the seventeenth century.

As human groups spread across the world, experiencing environmental and social challenges, our stories guided and bound us – and as our societies grew in complexity, our stories have evolved and adapted in concert, giving us the mental technology to navigate our physical and social environment as it expands from our immediate surroundings to the globalized world. We still use well-known stories – reduced to cultural maxims – to guide us, warning of 'the boy who cried wolf' or to 'look before you leap'. Storymaps, too, may have been widely used in the past – it has been proposed that Homer's *Odyssey* is a poetic, easily remembered map of the Mediterranean.[7] And there are hints that elephants use elements of storymapping, too. Like those of humans, elephant brains are oversized for their bodies, and evolution has favoured those individuals that are most adept at communicating and cooperating with enhanced memory. The matriarchs of the herds, like human grandmothers, remember remote water holes from long-ago droughts that will save the rest.

Stories are a powerful survival adaptation because they don't just allow us to travel back in time with our memories, they also allow us to mentally explore different future scenarios without expending time and energy. They act as virtual-world thought-experiments that enable us to trial risky or difficult permutations and store the outcomes. We do this intuitively all the time: we can imagine travelling the route to two different water sources and weigh up which is the better option without needing to make both journeys in person.

If we are told, 'Don't go near the boulders, it's dangerous,' we're less likely to remember – and so, survive – than if we're told, 'My cousin was sitting by those boulders and his face was eaten off by a lion that sleeps there.' Stories work as a cultural memory bank because the narrative device provides contextual 'infrastructure' that helps us understand, organize, share and store factual information.

Information told through stories is far more memorable – 22 times more, according to one study[8] – because multiple parts of the brain are activated for narratives. A list of facts only activates the language-processing areas of the brain (Broca's area and Wernicke's area, where words are ascribed meaning). However, the same information conveyed through a story also activates the brain areas relevant to the narrative: if the story involves jumping or running, the motor cortex lights up; whereas if it mentions someone's satin blouse, the sensory part of the brain is activated. Our brains react as though we were living the story and experiencing it firsthand. In this way, a storyteller can implant emotions, thoughts and ideas into the minds of the audience, making them feel as though they are experiencing the same events. In fact, scans show that the storyteller and the listeners' brains actually start to synchronize during storytelling – neurologists describe it as 'speaker-listener neural coupling'.[9]

In other words, our brains have evolved to understand the world through narrative, making stories phenomenally powerful cultural tools – another mutually reinforcing gene-culture coevolution. We weave narrative around all of life's events, we make sense of the world and our own lives through stories, and many of us give authorship of this ongoing saga to supernatural creators.

This is a quirk of our brain's sophisticated prediction system, honed by evolutionary processes for our survival. A big part of what our brain does is to receive sensory input from the rest of the body – our eyes, ears, skin, internal organs, and so on – and from this information it creates our perception of reality, our sense of self and our understanding of the world around us. We often describe this as consciousness. The brain is constantly updating its prediction tool with new sensory information, using its predictions to guide our interactions with our environment and enabling us to feed ourselves and avoid dangers, among other things. Our prediction system equips us[10] to expect weighty objects to fall downward; that objects in shadow appear darker; and that liquids don't need to be chewed, for example.

The brain creates a narrative to frame its snippets of received information, attributing agency to actors and seeking patterns, in order to make sense of happenings. A half-eaten cow and a roar can safely be attributed to a lion attack, so with this story in mind, you might prevent further losses by protecting the rest of your cows with an enclosure. If the cause is less obvious, if your cows die seemingly for no reason, we generate other narratives: it could be bad luck or it could be that the old lady in the village bewitched them, or that the spirits are angry. We can't do a lot to control the vagaries of luck, but we can drown the old lady or appease the spirits with offerings. If the rest of the cows survive after our actions, we update our story: it was because we've got rid of the old lady, or because the spirits are pleased with their gifts, or that our luck has changed. In this way, we add useful but also problematic beliefs to our cultural knowledge bank.

Humans are also inclined to see narratives where there are none because it can afford meaning to our lives, a form of existential problem solving. In a 1944 study[11] in the United States, 34 college students were shown a short animation in which two triangles and a circle moved across the screen and a rectangle remained stationary at the side. When asked what they saw, 33 of the 34 students anthropomorphized the shapes and created a narrative: The circle was 'worried', the 'little triangle' was an 'innocent young thing', the big triangle was 'blinded by rage and frustration'. Only one student recorded that all he saw were geometric shapes on a screen.

Since our brains essentially hallucinate for us a perception of the world around us,[12] it only takes a few tweaks to the input data to alter that hallucinated reality. This is such a powerful mechanism that we can change not just the story we tell ourselves about external events, like dead cows, but our own body's experiences. That's because the brain uses stories to help us interpret and respond to the sensory data it receives from our physical body, as well. If someone in pain is given a tablet by a doctor, who says it will ease the pain, it is likely the tablet will have that effect. The pain may go because, as the tablet is metabolized, it reduces histamine in the body. Or because the brain has directed the body to reduce histamine production because it expects the tablet to work. The story we tell ourselves about the tablet and the doctor can be enough to produce the tablet's biochemical response even if the tablet is a placebo – just made of sugar.

In fact, even when a person knows that the tablet is a placebo, its symbolic power may still be enough to generate a story that prompts the brain to produce the healing response.[13] The effect can be heightened by making the story more powerful – by the practitioner wearing a white coat, or the pills being packaged with pharmaceutical-style instructions and ingredients (some will list the chemical constituents of air, for example), or through culturally appropriate rituals. In some cases, a placebo delivered via a needle will have a greater measurable effect than one delivered by tablet, because the story is more believable that way.

Placebos work because the story is embedded in a person's cultural developing bath, and they work differently depending on that culture. This is behind the finding that placebo treatments for ulcers are twice as effective in Germany as in neighbouring Denmark and the Netherlands, whereas placebos to lower blood pressure are far less effective in Germany than in other countries.[14] Brain chemicals triggered by our beliefs can alter our responses to a range of triggers, including inflammation and stress. Chinese believers in traditional medicine and astrology, in which birth year is associated with a specific bodily organ that will eventually cause your death, die on average four to five years earlier of diseases of those organs.[15] This was such a startling finding that researchers went on to compare the mortality rates of Americans of Chinese and of European descent who had been born in the same disease-birth-year combinations – and found it held true. Chinese Americans believed this fatalistic story and so they made it come true – they were more likely to die when they had a certain disease, thereby adding credence to the cultural belief for other social learners. Longevity was determined not by genes but by the strength of a cultural story.

The power of stories to convince the brain to heal our bodies can also act in other ways. Epidemics of fainting and mass hysteria among teenagers and young women have been reported many times through history, with no clear cause. One episode that struck girls and teachers at Bibi Hajerah High School in Taluqan District, northern Afghanistan, in 2012, hospitalizing them, was blamed initially on a poison attack by the Taliban. After hundreds of blood and urine tests came back clear, the World Health Organization concluded the cause was a 'mass psychogenic illness'.[16] A similar event in the West Bank was blamed alternately on Israelis and Palestinians, until medics concluded that it, too, had a psychological cause. The hysterical contagion that sparked the Salem witch trials in Massachusetts was probably another example. In all these events the victims were experiencing fearful environments and their brains interpreted the story of their imminent danger by producing real bodily symptoms. Around 60 per cent of patients who are about to start chemotherapy experience anticipatory nausea,[17] because their brains have been conditioned to expect this side effect of the treatment.

It's called the 'nocebo' effect and it is the harmful opposite of the placebo.[18] Nocebos explain the power of curses, evil spells and black magic. Some even die of their curse. In one documented case, some 80 years ago in Alabama, a man who had been cursed with voodoo had become emaciated and was close to death by the time he was seen by a physician, Drayton Doherty. Nothing Doherty could say would shift the man's belief that he was about to die – a belief that his own brain was conspiring to hasten. Eventually, Doherty resorted to competing against the voodoo with another story. He gave the man a strong emetic, causing him to vomit, and through sleight of hand, produced a live lizard (from his bag). Doherty assured the man that the voodoo curse had caused this lizard to hatch inside his body, and now that it was removed, he would get well again. And so he did.[19]

From an evolutionary perspective, it makes sense for us to produce these bodily reactions in response to visceral feelings. If we are in a place of danger, if we have eaten food that is unsafe, then vomiting or fainting can act as warning signals to prompt us to flee or take action. Likewise, if we are in a place of safety and comfort, our brains accept this story as one to reduce pain and inflammation. This is particularly true of children, for whom the pain of a grazed knee can be assuaged simply by a parental kiss. It may also be part of the brain's strategy to align reality (the story it believes) with its sensory experiences.

Stories are the cognitive tools we have evolved in order to understand and interact with the world. We dream in stories and our inner voice provides the narrative to our waking hours, making sense of the world through stories in which we are the heroines, history is our warm-up act and the rest of the universe is the backdrop to our lives. Many of us understand our lives as 'a journey' with our purpose being a 'destination', and we may be 'lost' or 'at a crossroads' in our lives. Storytelling is a universal human trait. It emerges spontaneously in childhood, and exists in all cultures. We tell stories before we can talk, using mime or gestures: when my toddler shows me a butterfly and flaps her fingers in delight, she is telling me a story. Storytelling attaches emotions to events and, in part because of this, it makes stories memorable.

Our ancestors' immersion in storytelling can be found in ochre-painted caves and rock faces, dating back hundreds of thousands of years. Into a barren, unadorned earthscape, handprints and other deliberate paint marks were made by humans trying to communicate something beyond territorial claim. They speak to me of the human need to tell our personal story – to be known by another person. As author Kazuo Ishiguro put it: 'Stories are about one person saying to another: This is the way it feels to me; does it also feel this way to you?'[20] Ochre handprints have been found from southern Africa to Australia to Europe, an unbroken practice of human storytelling from our earliest days that, perhaps, predates spoken language. In 2017, the glass front of the New South Wales Supreme Court was defaced with ochre handprints as Aboriginal Australians protested against the lenient sentence handed to a man who had killed an indigenous boy. The red ochre demanded justice and told the story of a link back through time to when the continent's first people used it as a cultural tool.

Deep in northern Spain's Basque country, in Cantabria, there is a cave complex where three valleys meet between two tributaries. Perhaps because it was located on a natural migratory route for herds of game, El Castillo cave was for thousands of years home to Neanderthals and then our own ancestors, offering refuge during the bitter ice ages. The walls of the cave's labyrinthine rooms are covered in extraordinary paintings made variously by these two human races from 64,000 years ago. However, it is in a chamber deep within that something so unlikely has been discovered that scientists have only recently been able to figure it out. For my visit, I asked my guide to switch off the lights – I wanted to view the spectacle as it was intended.

My eyes strained into the darkness for a few seconds, then, with a flick of the guide's torchlight, an animated, three-dimensional bison-man monster appeared, looming terrifyingly down on me from the ceiling. As the torch moved around an imposing three-metre-tall stalagmite, the man-beast grew and contorted, then walked across the ceiling. A primal feeling rose inside me, a mixture of reverence, wonder and fear. This incredible imagery was a form of prehistoric cinema. At least 15,000 years ago, an animator captivated their audience by moving a fat-burning lamp, cleverly using the bulges of the stalagmite to cast light and shadows onto different painted cartoons in a series, evoking movement and meaning. A story told this way would have infected the imaginations of others with an idea that was invented in the creator's mind. Storytelling provides unparalleled opportunities for social bonding. The lies we tell in stories are consensual: we agree as an audience to enter that liminal space, to cross the threshold into a new reality, an imagined landscape.

The multi-sensory experience of cinema exaggerates this effect. Part of its power is the scale – close-ups used in modern movies make a huge difference to how the brain perceives a face or human form image. (And animating at 12 drawings – or 24 frames – per second, there is no time to rationalize and our brains have difficulty telling us what is real and what is not, and whether we know someone or not, so the characters in a movie feel more intimately connected to us.) It is entirely clear that the creators of the ancient cave cinema knew the feelings they were awakening. There are unambiguous drawings of the bison-man – perhaps representing a shaman cloaked in the hide of a bison – on the contours of the walls and stalagmite in his various apparitions. What worlds were created in this mysterious dark cave? What shamanistic hallucinations bound them together in shared enterprise and belief?

The many mysteries to life, the things we struggle to explain through demonstrable examples, we reason into acceptance through stories of imagined gods and magical powers. For many people, the distinction between the invented and observable worlds is unclear and even unnecessary. Such stories can bring us comfort – for a highly socially dependent species, gods become the ultimate social support in the face of adversity. Religiosity increases after earthquakes, for instance.[21] Praying to a caring god can reduce stress levels and provide reassuring rituals and the social support of a community that persuade the brain to reduce the body's pain response,[22] for example. Religious people are less anxious about making mistakes,[23] perhaps because religions involve a degree of fatalism and deities with ultimate responsibility, which buffers against second-guessing decisions and may have helped us survive, pointing to an evolutionary selection pressure.

Indeed, seemingly irrational story-based practices may be so widespread because they are actually advantageous. Take hunting: societies around the world cloak their hunts in rituals, including imitating the animals, only hunting in certain areas and pursuing directions that would appear to make for less efficient meat gathering. When researchers analysed the success of ritual-directed hunts, they found these often worked out to be a better strategy than the rational method of seeking patterns in successes and preferentially copying those. The problem with pattern-seeking when selecting hunting sites – returning to locations where prey were successfully hunted before – is that the prey learn to avoid those locations. Using storybound rituals randomizes the hunting areas and helps hunters to avoid biases, which are the Achilles heel of human cognition. Chimps, for example, don't suffer from biases and are better able to randomize their efforts.

Stories also provide a mechanism for ensuring shared natural resources are used sustainably and cared for by the whole group – it is no surprise that animism is widespread among hunter-gatherer societies, and may have been present in early hominins, predating language.[24] Most such religious stories hold humanity in a reciprocal relationship with the natural world (the Judeo-Christian idea of humans having dominion over nature is unusual). The Yakut indigenous people of Yakutia in Siberia, for example, hunt reindeer but they believe that the agency resides with the deer, who give themselves up to be harvested. Every kill is done with ceremony and includes recognition and deference to the animal's spirit and its gift to the human tribe.

Ancestors play an important role in these environmental belief systems, and in many cultures, ancestral spirits are thought to reside in other animals or natural forms. When someone dies, they often continue to play a role in the living community, strengthening the social ties across generations and across life forms. Death customs are a part of every culture's stories and many of the most important decorative items archaeologists have recovered were used to adorn the dead. Cumulative cultural evolution is only possible if there is an unbroken line of cultural transmission over generations – even when individuals die, the cultural practices they helped to underpin must survive. Stories and rituals that include ancestors and allow them to remain part of the group's narrative world help support this unbroken transmission and strengthen social bonds, which is perhaps why they are so prevalent. We still use material objects to create cultural memories and narratives of our dead, from cenotaphs to Marilyn Monroe posters.

People who spin narratives for a living are celebrated throughout the world. The Agta, a Filipino hunter-gatherer population, value storytelling more than any other skill – twice as much as hunting ability – and the best storytellers have the most children, anthropologists report.[25]

Stories put listeners on an equal emotional wavelength, eliciting understanding, trust and sympathy. From humanity's earliest days, fire has been an important part of this process, extending the day and allowing imaginative conversation to flourish. Anthropologists analysed the conversations of modern hunter-gatherers in Namibia and Botswana, and found that while daytime discussions exclusively deal with mundane matters of economic issues, land rights, and so on, more than 80 per cent of firelight conversation is devoted to telling stories.[26] We interpret the world, invent our own unnatural worlds and beasts, and we convey these ideas to others through storytelling, art, song and dance: mind speaking to mind. These communal rituals are potent, strengthening bonds and cementing trust and solidarity. From football chants to religious hymns, the brain processes the experience of singing[27] and dancing in concert with others as though we share much more than simply those few minutes of joint activity[28] – we become a pseudo family through it. Experiments show that after singing and dancing together as a group, individuals are more cooperative and will contribute more money into a shared, group purse, resulting in better payouts for everyone.[29]

It is not enough for our brain's prediction system to generate its own narrative: we have to ensure our individual narrative is aligned with that of our group. Stories have the power to bind us together in shared belief and play a key role in bringing together strangers over shared enterprise. So, although telling stories does not contribute food or other tangible resources to a group, it is likely that storytelling skills evolved as an adaptation for group cohesion and cooperation, and for reinforcing social rules and imparting cultural knowledge. Among the Agta, for instance, anthropologists found that groups with good storytellers showed greater levels of cooperation and sharing.[30] Around 80 per cent of stories told by the Agta convey messages about cooperation, sexual equality, egalitarianism and punishment of rule breakers, all cultural behaviours that enhance group survival. Those societies whose stories had fewer cooperative messages – they were more about nature, for example – cooperated less as a group.

Stories make our societies more cooperative and help us to cooperate as individuals. We pass around information about ourselves, other people and our world through stories, learning how to relate to people, how to empathize, how to behave. Through stories we can explore the human condition and see how other people think. This can affirm our own beliefs and perceptions, but it also challenges them. Regardless of the language, there is something universal about what occurs in the brain at the point when we are processing narratives, triggering better self-awareness and empathy for others. Psychologists scanned people listening to narratives in English, Farsi and Mandarin, and found the same patterns of brain activation when people found

meaning in the stories.[31] Other studies find that reading fiction significantly increases empathy toward others, including people of a different
race or religion. And the more absorbed in a story a reader is, the more
empathetically they behave in real life. For instance, if the researcher
'accidentally' dropped his pens, those participants who had previously
reported being 'highly absorbed' in the story were about twice as
likely to help pick up the pens.[32] Another study concluded that literary
fiction 'uniquely engages the psychological processes needed to gain
access to characters' subjective experiences'.[33] That's to say, if you
read novels, you can probably read emotions, vital skills for forming
cooperative societies.

Stories are also a useful way of presenting novel ideas or behaviours to people who may be resistant to them in real life, thereby
easing the cultural evolution of different societies and institutions.
The collective nature of stories makes them harder to destroy or control because the information is distributed. In this way, subversive
messages survive to strengthen disempowered groups.[34] In deeply
conservative Afghanistan, two-line anonymous poems called landays
are created and passed orally between Pashtun women, in which they
tell their own forbidden stories of sexuality and female liberation:
'When sisters sit together, they're always praising their brothers /
When brothers sit together, they're selling their sisters to others' or
'Embrace me in your suicide vest / But don't say I won't give you a
kiss.'[35] Storytelling allows people to try out dangerous political and
social ideas, such as the emancipation of women or slaves, which
leads to real-world change. Indeed, books can be remarkably influential: George Orwell's 1984 and Mary Shelley's Frankenstein are
continually referenced today; the Tuscan poet Dante Alighieri is cred-
ited with Italian becoming a national language after writing his
Comedy in that local tongue rather than Latin; and Alexander the
Great used Homer's Iliad as a blueprint for his own conquests, reportedly sleeping with a copy of the epic at all times.

Epic narratives help form national identities, telling audiences
where they come from, who they are and how to regard their neighbours. The replication of stories forms a shared history that glues
society together – in many languages, the word for 'story' is the same
as that for 'history'. It is through stories that we develop and share

our invented notions of democracy, patriotism and other ideologies. Fairy tales, for instance, are generated from a human desire to transform the world to our needs and to impart lessons. Some specific European fairy stories, such as *Beauty and the Beast*, have roots that stretch back for around 6,000 years, to our common Indo-European ancestry, literary anthropologists have discovered.[36] Tracking the ancestry of these stories reveals the enduring signatures of ancient population expansions and dispersals, and the extraordinary strength of story transmission over millennia – the lesson that a kind heart can reside in an ugly person is evergreen. It's why Europeans still regularly transmit the lessons of Aesop's fables, which the Greek slave invented some 2,500 years ago.[37]

It seems we have been telling the same stories for millennia, updating the characters and details for different audiences and times. In 1872, when George Smith cracked the complex triangular cuneiform code on a series of Babylonian clay tablets, he revived for us the oldest written story in the world. But the Epic of Gilgamesh, an extraordinary 4,000-year-old poem of romantic drama, adventure and a quest for immortality, turned out to be strangely familiar. On the so-called 'deluge tablet' of Gilgamesh, a character named Utnapishtim is told by the Sumerian god Enki to abandon his worldly possessions and build a boat. He is told to bring his wife, his family, the craftsmen in his village, baby animals and foodstuffs. It is almost the same story as, and surely the inspiration for, Noah's Ark in Jewish, Christian and Islamic texts.

Indeed, around the same time as Gilgamesh was being pressed with reeds into wet clay, an Egyptian scribe called Ankhu was lamenting that there was nothing left to say that had not already been said: 'If only I had unknown utterances . . . free from repetition, without a verse of worn-out speech spoken by the ancestors!'[38] There may be only a handful of basic plots but from these limited rules, we weave myriad possibilities. We don't even need to invent new stories; the ones we have simply evolve adaptations to their new listener environment. We can always say it differently and to different people.

We generate the stories we need, and these reflect the cultural developing bath of their times, providing an interesting window into how this changes. Initially, many religious stories were unconcerned with moral codes and policing human behaviour. The gods of many of the first documented religions led exciting, soap opera lives with power over us – appeasing them involved rituals and sacrifices, which would sometimes be rewarded by divine help. Shame, though, was an important motivator. In *The Iliad*, Zeus is uninterested in justice. The ancient Greece of the time was a family-based patriarchy, where even adult sons were denied any rights until their fathers died.

By the time of *The Odyssey*, perhaps 50 years later, things had changed. It was a turbulent period with invasions, economic crises, class warfare, social upheaval and personal insecurity. The clan system began to weaken and a growing clamour for individual rights and personal responsibilities was challenging the powerful moralistic family patriarchal unit. The Greeks seem to have projected their own demands for social justice onto the cosmos. Zeus of *The Odyssey* was much more judgemental, complaining that men 'by their own wicked acts incur more trouble than they need'. And once Zeus became moralistic, he lost his humanity, and Olympianism became a religion of fear. There is no word for 'god-fearing' in *The Iliad*, whereas in *The Odyssey*, to be god-fearing is an important virtue worthy of praise.[39] There's another change in the air, too. The perhaps universal fear of pollution (of miasma) and ritual purification (catharsis) increases. In *The Iliad*, when people carried out fairly tokenistic purification rituals, the characters breathe clean air. In later versions of *The Odyssey*, there are demons polluting the air and Oedipus becomes a polluted outcast. Pollution begins as an external, indifferent event, like a germ that randomly infects people, but it is also hereditary, passed down through families, shaming each generation until it is cleansed. From here, miasma evolved into the idea of sin, that it's a disease of the will, and people feared falling into sin. Cathartic rituals started to be more complex and involved cleansing the mind as well.

Stories are an incredibly powerful cognitive technology because it is through them that we invent ideas, such as sin, that we then collectively believe. These go on to shape our behaviour and societies, influencing our reproductive success and deciding survival, such as through executions or abortions. In this way, our cultural inventions can become drivers of biological evolution, telling us who it is sinful to share our genes with, for example.

So, storytelling is an adaptation that prolongs the life of our ideas

and inventions, enveloping cultural information for faithful transmission between people. But, as our societies grow larger, it becomes
important to store non-narrative data, too, such as who owes what to
whom. This was achieved through physical, visual records, from the
intricately knotted threads used by the Inca to scratched shells, to
scored clay, or stone tablets. In Australia, people have used 'message
sticks' for tens of thousands of years, to transmit information, including invitations, trade negotiations and requests, across the massive
continent. These foot-long wooden sticks were inscribed with symbols that could be understood by people in different regions, and also
acted as diplomatic passes through other territories.[40]

Around 5,000 years ago, humans invented a most brilliant and flexible information-storage tool: writing. This was by far the most
energy- and time-efficient way of managing, storing and transmitting large amounts of information with high fidelity, which, as we
have seen, is key for cumulative cultural evolution.

However, the investment in time (and child labour) that learning to
read and write requires is considerable, so it has only been adopted by
societies where the payoff is worth it. For hunter-gatherers living in
small, multilingual populations spread over a large area, there is not
sufficient selection pressure to adopt writing. The idea of 'property' –
land, bushels of wheat, numbers of goats and children – for which
accounts might need to be kept evolved after people settled. Even for
many agrarian societies, the type of crop may be key. Cereals with a
regular harvest, like wheat or rice, are much easier to tax, enabling a
state to develop its infrastructure to the point where writing becomes
advantageous or necessary. Even so, only a small section of a rural
population may be literate, usually men in roles such as government
officials or religious leaders.

The societies that developed and used writing were settled ones
that had managed to produce a surplus of food to support a large population with diverse trades, with power and control over a large number
of different clans, and secure and stable enough that they weren't
constantly at war. Around 3000 BCE the first crowded cities and
states grew up, supported by the cereal farmers of Mesopotamia.
This dramatic societal change from broadly clan-based groups to a

far larger state of anonymous individuals was seismic, and writing was one of the tools that enabled it.

For the majority of human existence, there have been no written records of what people have said and done. History began once people permanently inscribed into clay the most mundane of accounts: property ownership for taxation or trade, the inflows and outflows of commodities from a city's ports, the wealth of rulers and their change-able rules of law, lists of battle victories. From these early Sumerian scribbles to our modern Facebook accounts, our compulsion to record our lives has proved irresistible. This evolution in information storage and transfer enabled societies to grow in size and complexity to become concentrated, networked hubs of cultural knowledge.

Writing that could convey more complex data (using symbols to represent four cows, say, rather than pictures of four cows) and then writing that could convey true speech followed in time, and evolved separately in different cultures. This crucial step involved agreeing a set number of visible marks to represent spoken sounds – an awesome achievement successfully made by countless societies, many of which adapted each other's symbols for their own writing systems. They range from ancient Chinese script to the pared-down brilliance of the alphabet, with (roughly) one symbol for one sound. The alphabet[41] was invented only once, and was, according to the ancient Greeks, the greatest gift – greater than the gift of fire – to humanity by Prometheus. The alphabet itself is based on an early Semitic script. *Alpha*, *beta*, and so on mean nothing in Greek, but the α (when transposed on its side) represents the horns of an ox – *aleph* is the Phoenician for 'ox', descended from the Canaanite 'alp'; whereas β (on its side) represents houses with curved roofs – *bet* is the Phoenician for 'house' (you can see it in 'Bethlehem', for example), and its likely origin is an Egyptian hieroglyph for 'house'. This Phoenician invention is the ancestor of the many and diverse alphabets we use today, from Arabic to Latin.[42]

The alphabet is continually evolving – English has lost six letters over recent centuries, including ð ('eth', the hard version of *th* used in 'the'), þ (or 'thorn', the soft version of *th* used in 'thing'), and ȝ or ('yogh', the throaty *ch* of 'loch').

To us now, living in a world so utterly infused with the written word, it seems barely possible that large, urban societies functioned without it. However, just as fish that become marooned in caves for many generations evolve blindness once sight becomes redundant, so too cultures can lose their technologies and practices – sometimes for centuries.[43] It's another reminder that cultural evolution has no direction; we are not necessarily 'progressing' toward something better. One such 'dark age' occurred in ancient Greece following a series of devastating invasions and natural disasters. By 1200 BCE, Greeks were living in the ruins of their former civilization, no longer able to read or write.

It is, then, extraordinary to realize that it was during this dark age of illiteracy, perhaps in the still-important port city of Izmir (Smyrna), that Homer created his immortal epic poems. Poetry is composed, like music, to be performed. The words, the metaphors, rhythm and musicality come alive when spoken aloud. Homer, the legendary blind poet, would perform his poems from memory, and his audience would themselves memorize them to recite to others.[44] But even though the poet and his contemporaries were illiterate, they knew of writing. They were surrounded by the ruins of inscribed temples and monuments, and they traded with literate societies, including the Phoenicians.[45] Homer himself makes tantalizing reference to the art of writing in *The Iliad,* when a messenger carries a folded metal tablet inside which is written, 'Kill the bearer of this letter.'

Imagine being a story maker in a time when writing had been lost, but knowing of its existence elsewhere. Yet there was no better time to be a blind author. Homer and his contemporaries relied on another cognitive technology: people in non-literate societies can memorize better.[46] The wonder I feel at the Homeric era's loss of writing would probably be matched by their surprise at our memory loss. Epics like *The Odyssey* were composed to be easily memorized – they had a strict meter, which helped with remembering lines and improvising, and included plenty of repetition, so common phrases could be slotted in like choruses. Nevertheless, remembering thousands of lines of verse, as educated people could, involved skill, and would have changed their brains in a measurable way, rather as modern-day London taxi drivers, who must memorize thousands of street names and directions, develop structural changes to their brains,[47] such as an enlarged hippocampus.

The Greeks invented a sophisticated art of memory called mnemonics, a culturally learned technique that works rather like Aboriginal songlines that mentally fix narratives to landscape and constellations. The story goes that the Greek poet Simonides of Ceos was booked at a banquet to recite his work. Just after he left the building, the roof collapsed, killing everyone inside. Their bodies were utterly disfigured but, by using the mnemonic technique, Simonides was able to travel around the hall in his mind and remember where each of the guests had been sitting, allowing them to be identified for burial. He reportedly went on to develop his technique as a way of impressing memories into an imagined 'mind palace'.[48] This works because it uses our culturally and biologically coevolved storymapping ability to conjure an architectural space and fill it with things. Travelling around this mental palace subsequently – reliving the story – is an extremely effective way of remembering long lists of information, a public speech, or, indeed, an epic poem.

It is, however, cognitively demanding, and literacy allowed us to outsource these energy costs and rely on the external collective memories held in libraries and, more recently, online.

Learning to read and write, like most culturally acquired skills, changes our biology (although not our genetics). Literate people have different brains from illiterate people by around age eight, because their visual processing systems have been specialized for reading. Some of these modifications improve the networking between different sides of the brain, enhancing object recognition and verbal skills but reducing cognitive abilities in other areas, such as facial recognition. Highly literate people become word spotters in the same way that a hunter-gatherer is able to detect the nuances that reveal animal tracks. Our eyes jump to word patterns in our native script and we unconsciously decipher them, compulsively reading across our environment.

Even when words or letters are jumbled up, it doesn't disrupt our reading too badly – yuor biran aumtoacitally flls th gpas.[49] Our brains are good at reconstructing writing (and speech) using context. That's partly because skilled adult readers don't use overt translation of reading into sound (as children do), but a direct pathway from letter patterns onto meanings that's highly efficient and very rapid. The average adult reads at about 230 words a minute and recognizes about

42,000 words by the age of 20. After this, people typically learn one or two new words each day, so people in retirement have vocabularies quite a lot bigger than recent graduates. In this way, the knowledge accumulated by elders makes them important repositories of cultural richness and diversity in human societies.

The physical act of writing also uses multiple brain regions and has wide-ranging cognitive effects. Writing something down stores the information on paper and also in your memory, because it stimulates a collection of cells in the base of the brain that filter information and focuses your brain's attention. Writing organizes our thoughts and makes elusive, nebulous feelings crystallize out onto the page, shaping our thoughts into a form that is comprehensible and sharable. It makes the unfathomable possible to see. *Text* is from the Latin 'texere', 'to weave', because we weave our words just as we weave textiles.

The invention of the printing press, the availability of cheap paper,[50] and a new literate class of burghers and traders democratized information, generating writers and readers from every sector of society. Now, from the age of 11, reading is the main way we learn new knowledge in literate societies. Writing is incredibly powerful because of its reach – I don't need to meet the author of the words I hear in my head, yet her words will appear in my mind as faithfully as if she had whispered them into my ear. It is no longer necessary to memorize information that can be quickly referenced. We learn instead where to go for information, and to discern from among the great avalanches of literature which ones have value, just as we learn which people to copy.

Books are more reliable and keep cultural information in societies for longer than oral stories and provide another mechanism for cumulative cultural evolution: books referencing other books build on their writers' knowledge. The stories in the Dead Sea Scrolls, which date to perhaps 250 BCE, are almost identical to those in the so-called Leningrad Codex, written around 1,000 years later; faithful copying by scribes had preserved their fidelity. What's more, they were based on oral stories that had been passed down generations over a further 1,000 years, at least since the time of King David, before Hebrew was a written language.

The invention of writing did more than improve the way we store and disseminate information, though. It fundamentally changed our encultured collective minds, extending and outsourcing humanity's processing power. This catalytic boost took our societies and technologies to the next level of complexity. For philosophical arguments, logical reasoning, abstraction and higher mathematics to develop and benefit from the input of multiple thinkers, they need to be written down and worked out on a page so that the idea visibly progresses from the one before and enables the next one, and can be scanned and analysed in an entirely different way to verbal arguments. And this allows more complex social entities to be developed, such as governments, a civil service and economies based on money. That's how developments in writing led to developments in human organization.

Paper is still widely used despite its much-anticipated demise. But information is now being stored in digital bits based not on the digital categorizations of phonemes or word sounds (with which we organize our speech), nor on the digital categorizations of the alphabet (with which we organize our writing), but in binary ones and zeros to be stored in silicon chips. In this sense, information itself has physical properties like energy and matter; manipulating, storing and transmitting information costs energy, and 'forgetting' information – wiping a disk – is expensive and difficult to do. In the coming decades, we will be using the ultimate biologically evolved information storage system to store our data: DNA.[51] The structure of DNA encodes the genetic information used to build the proteins of life; it is this biological system that generated cultural beings with the vision, creativity and technological knowledge to use the stuff of our being to store the stuff of our thinking.

Our invention of stories provided a collective memory bank for our accumulating knowledge, improved the fidelity and reach of cultural transmission, and bound our societies into closer cooperation. In these ways, stories reduced the energy costs of cultural evolution and improved our survival. Storytelling and our reflexive use of narrative became a biologically evolved part of our cognition that shaped our minds, our societies and our interaction with the environment. Language is the currency of stories, and we shall investigate this next.

7

Language

*High in the rocky mountains of La Gomera, in the Canary Islands,
a duet is playing out. The rugged steep-sided cliffs of this volcanic
island are cut by deep ravines and separated by wide valleys, and yet,
from far, far away, clear notes pierce the subtropical air. I wait in
silence, listening over birdsong and the occasional fussing and bleat-
ing of goats picking their way over stones. Then, from somewhere
just above me, comes the tuneful reply.*

*The people here communicate across the unforgiving landscape in
an ancient whistling language called Silbo that carries their conversation as far as eight kilometres, from mountain to mountain,
between remote farms and villages. As one old goatherd says, it's
cheaper and faster than using a mobile phone, and it never lacks signal. Silbo is now taught in La Gomera's schools, although many
children learn it along with Spanish as a mother tongue, inserting a
knuckle into their mouth to make the sounds, or learning to perform
particular tongue folds. It is so like birdsong that blackbirds have
been known to mimic dialogue.*

Communication is a fundamental characteristic of being alive, and
every life form makes itself known through some sort of signalling.
Plants message each other via networks of soil fungi, and cephalo-
pods use skin colour. Some mammals, such as dolphins, apes and
dogs, communicate so proficiently with humans that they've been
described as having primitive language. However, human language,
whatever its medium, requires a level of comprehension not seen in
other animals. Chimps can be taught to whistle but they show no
musicality,[1] and they have no language. There is a yawning chasm

between their communicative abilities and ours. Chimp-talk, for instance, only has five basic sounds; all chimp calls are context-specific, and, unlike humans, they will never use calls out of context – a chimp will never use the predator call if there's no predator. What we humans invented is a truly flexible communication tool with rules.

Language is more than a system of transmitting information, though. Language is what makes us human in a fundamental way. Words are thoughts. Without language we have no inner monologue, no system to arrange or formulate our thoughts. The feelings we notice are the ones we label. People with aphasia (loss of language, usually through stroke or brain injury) are no longer able to mentally time-travel, to recognize associations between things, or to follow an argument. They are stuck in the literal present and struggle with the most basic human thought processes. I have language, therefore I am.

In the same way that Earth's many different environments have driven genetic evolution, so environmental pressures have also guided the cultural evolution of language. Different dialects and languages are often separated by geographical barriers and are influenced by the local topography and its acoustics.

Whistling languages evolved as an adaptation to steep terrain, dense forest or ocean, where it might otherwise be hard to communicate at a distance,[2] because whistles carry much farther than normal speech and tend not to startle prey as much. The first people to arrive in La Gomera, some 7,000 years ago, may have brought their whistling language with them from the Atlas Mountains in North Africa, where Berbers still speak a whistled Tamazight language. It has proved very useful in the past for passing clandestine messages during their resistance against French occupation. Similarly, during the Second World War, the Australian army recruited Wam speakers from Papua New Guinea to whistle messages across the radio to confound Japanese eavesdroppers.[3] Around 70 different groups are known to speak in whistles, including hunter-gatherers in the Amazonian rainforest, Arctic Inuit whaling communities and Greek islanders. Hmong communities of the Himalaya also chat in whistles across forests and farms, and when courting to anonymize flirtatious exchanges between houses (it is harder to identify a whistler than a speaker).

There are parallels in the animal world. It's been known for a few decades that in woodland areas, where trees muffle or distort sound, birds tend to sing songs made up of lower frequencies, with less variation, than those living in open areas.[4] Biologists have recently noticed that some city birds are adapting their songs the better to be heard against the background din of urban life, singing lower frequency songs with simpler structures than relatives living in quieter places. Now, scientists are finding the same kind of adaptation among human languages. The number of consonants in a language, and how consonants cluster together in syllables, seems to depend on the average annual temperature and rainfall, the amount of vegetation and the elevation and ruggedness of the place where the language is traditionally spoken.[5]

Languages spoken in warm, wet and heavily wooded areas, such as in Southeast Asia, tend to use more vowels and fewer consonants, mostly in simple syllables. By contrast English and Georgian languages, which did not evolve in rainforests, burst with consonants. The languages of those who live at altitude contain more words that have a strong expulsion of air in the consonants. Meanwhile, arid, desert-like places are less likely to have tonal languages like Mandarin or Vietnamese, in part because of the harmful effects of dryness on vocal cord movement – an anatomical, environmental cultural adaptation.

Spoken words are essentially a series of sounds, ranging from high-frequency consonants, like f, p or t, to low-frequency vowels, like e, o and u. Obstacles, such as dense vegetation, or heat rippling the air, act as selection pressures on language, because they cause high-frequency sound waves to distort or become lost. So the differences in languages are, in part, cultural adaptations to different environments.

The whole human evolutionary triad is affected, because these acoustic variants are also driving our genetic evolution. There is evidence that the emergence of non-tonal languages – such as European ones – has, over the past 50,000 years, been influencing the spread of two new gene variants involved in brain growth and development.[6] Tone describes the use of pitch, timing and loudness to convey meaning – in a non-tonal language like English, tone changes the value of words, and helps listeners cope with long sentences by breaking speech into chunks; in a tonal language, it changes the actual meaning of the word or phrase. One sound, for example '/ma/' in Mandarin, can mean 'mama', 'horse', 'hemp' or 'scold', depending on how the tone is used. The Hmong language uses as many as eight tones to give different meanings to sounds. Some tonal languages have evolved into non-tonal ones, for example, the Greek used in Homer's time was tonal, whereas modern Greek is not.

For tonal languages, the nuances of the phonemes (the consonant and vowel sounds) are less crucial, so music, such as whistling, or even drumbeats, can carry conversations more easily. Sub-Saharan Africa was once crisscrossed by a communication network of drumming villages, in which everyone could understand the one-dimensional drum language. Villages along a route worked in relays to convey drummed messages, poetry, announcements, warnings, jokes and prayers across vast distances, so that complex announcements could be sent a hundred miles or more in an hour. Such efficient transmission wouldn't be achieved anywhere else until the invention of the telegraph.

Whistling and drumming languages force their speakers to combine language and melody processing in the brain, and could hold clues to the origins of speech.[7] Music and language[8] are both processed by the same brain regions[9] and appear to be linked in other ways, too – studies have shown that music classes improve literacy. Some linguists believe human speech began[10] with a musical proto-language such as whistling, which is within the physical capability of apes. The Hmong whistlers often substitute a mouth organ for whistling, communicating in a fully fledged musical language.

The evolution of a biological ability for language was directed by its cultural invention, and vice versa: the anatomical changes to our jaw just a few thousand years ago, after we adopted agriculture (softer foods allowed smaller jaws and a new overbite), meant we could make 'f' and 'v' sounds, and it spurred an explosion in new languages, linguists believe.[11] And yet, humanity's greatest invention was never truly invented – it evolved. Cultural evolution generated language in a not dissimilar way to how it generated cooking, and we are as dependent on it. Every human society has complex language. Using language is a biologically evolved instinct: even though we are not born knowing language, it must be learned from others, and yet it doesn't need to be taught. Language is a riddle of paradoxes – 'half-art, half-instinct', according to Darwin.

The neurological basis to all of this is unclear, for there is no 'language' centre in our brains. Instead the ability appears to be properly nebulous. In this way, language pervades the biology of our brains in the same way as it pervades our culture. We can speak within months of birth, and without any formal teaching other than listening to conversation. This remarkable ability is almost entirely universal, even in children with reduced intellectual abilities. Probably, this genetic talent coevolved with human babies being born so small, helpless and undeveloped that they must be closely looked after for many months.

How did an ape begin to talk? Some scholars believe our spoken language evolved from primate vocalizations; others point to the apes' repertoire of gestures, and think speech developed from that. Most likely, it emerged from a combination of both. Complex, rich sign languages were, until fairly recently, commonly used by Australian and North American hunter-gatherer societies. Plains Sign Talk, for example, was used for talking, stories and trade across large swathes of North America until European occupation. And gesture languages continue to be used by deaf people globally.

Even our most inane pronouncement involves a highly coordinated oral dance of such complexity that, if it had to be consciously thought through, we would all speak less and sound wiser. To produce our expansive repertoire of vocalizations, our ancestors evolved through a suite of anatomical adaptations, starting with bipedalism, which allowed for much better breath control by freeing the ribs and diaphragm (that previously supported the forelimbs) and opening up the vocal tract. Also important was the descent of the larynx (voice box) to an extension at the back of our tongues, suspended from a small, but critical, horseshoe-shaped bone called the hyoid. This allowed better acrobatics of the vocal tract and more space for the tongue to move during speech production, which enabled us to make vowel and consonant sounds. This was an evolutionarily risky strategy: our lowered larynx means we can no longer swallow and breathe simultaneously, and are far more likely to choke than other primates, whose larynxes are high up in the nasal cavity. Human babies are still born with a high larynx, like a snorkel, so they can breathe while suckling. But, at around three months of age, the human larynx descends, and the payoff is worth it: apes, with their higher larynx, can't speak as we do, even with training.

Every vocal performance involves hundreds of thousands of micro-collisions in the throat. Each utterance depends on a pair of thin, reedlike, muscular strips, the vocal cords, located inside the larynx. When we are silent, the cords remain apart to facilitate breathing. When we sing or speak, air is pushed up from the lungs, and the edges of the cords come together in a rapid chopping motion, creating sound. The greater the vibration, the higher the pitch. By the time a soprano hits those lush high notes, her vocal cords are thwacking together 1,000 times per second, transforming a burst of air from her lungs into music powerful enough to shatter glass.[12]

It's not clear when in human ancestry speech emerged but we could have conversed with Neanderthals.[13] They too had these voice box adaptations that are critical for speech. They also shared a very similar version of our FOXP2 gene, the so-called language gene. People with a mutation in this gene have problems learning to speak, pronouncing words and understanding and making sentences. FOXP2 exists in many other animals, but the human version, which is different from a chimp's by two DNA letters, is quite recent. This tiny tweak to two of the 740 bases in our FOXP2 gene appears to have been transformative.[14] We know it alters the expression of more than 100 other genes compared with the chimp version,[15] many of which have roles in brain development and function, and also affect soft-tissue formation and development, linking FOXP2 to both the cognitive and physical side of speech and articulation. When researchers put the human version of FOXP2 in mice, the rodents produced more frequent and complex alarm calls, and were better at learning to solve puzzles.[16] The multiple survival benefits of better communication and learning would have rapidly spread the new version of FOXP2 through the entire human population, and our cultural invention of language coevolved with it.

We are born with an innate ability to learn the rules of grammar and several thousand words – what Steven Pinker terms 'language instinct' – and strong desire to communicate. Bipedalism freed our hands, allowing us to gesticulate in ways that other animals cannot, and we trademarked pointing. It takes babies several months before they understand the relevance of pointing, but by 12 months, they are doing it themselves, beginning their first 'conversation'. Pointing is a surprisingly complex, uniquely human action, which requires a sophisticated understanding of what's going on in other people's heads – and, just as importantly, the curiosity to wonder in the first place. By pointing, a child is able to communicate something specific: he wants something – give me a banana (imperative pointing); he is explaining something, sharing information – that's the chair you can use; or to share an experience – look at that balloon (declarative pointing). This last – this meeting of the minds to share an opinion – stems from our innate desire to cooperate[17] and is at the root of the way we collaborate as a species.

Communication begins powerfully with the eyes – even from birth, a mother can influence her infant's gaze direction simply by moving her eyes. Ape mothers, by contrast, must turn their heads towards the object to show that there is something worth looking at. Humans evolved exaggerated white scleras to reveal who or what we are looking at, and we can detect an eye movement of just a degree from a couple of metres away (corresponding to about a five-centimetre shift in the point of attention, such as from the left eye to the right). Indeed, eye contact is such an important part of our social cognition – and sense of self – that young children struggle to understand that someone is even present without it. If you've ever wondered why preschoolers playing hide-and-seek simply cover their eyes to hide, the answer is that they require gaze reciprocity to 'see' another person.[18] They will also claim they can't hear someone whose ears are covered or speak to someone whose mouth is covered.[19]

Young children are acutely aware of the reciprocal nature of human communication, and have an innate tendency to acquire knowledge by joint attention. This means they undergo a developmental period in which they believe the self is something that must be mutually experienced for it to be perceived. In a 2003 experiment in which American infants were taught Mandarin in three groups (video, audio and flesh-and-blood teacher), only those with a human tutor learned anything at all. Shared attention is the starting point of conscious human learning. It is why infants don't learn to talk from video, audio or overhearing parental conversations. We haven't evolved to. We need reciprocity to be validated as separate people. When we use speech, we are not simply announcing information, like an audible robot or alarm clock: we are targeting another's mind and expecting a response, even if that is simply an acknowledgement that you have been heard. Our other emotive utterances – laughter and crying – are also strongly communicative. Indeed, laughter is very contagious, particularly if it's done by someone we know.

Language is another vital survival skill that we have evolved to be entirely reliant on others to acquire. There's a narrow window in our childhood when we can pick up language, and if we're not surrounded by other people conversing in that time, we won't ever speak like a true native. The process begins before birth – foetuses can recognize the sounds and rhythms of their mother's tongue and prefer it to other languages[20] – although it takes several years for children to unconsciously master grammar, vocabulary and the intricacies of muscle control and movement required for speech. As with other aspects of cultural learning, the cultural developing bath is crucial: the number of words a child hears by their third birthday strongly predicts academic success at age nine, and this variance is socially determined and stark. In one study, by the age of four, the richest children had heard 30 million more words than the poorest.[21]

However, the difference in language skills is not simply down to hearing more words. A more recent study on four-, five- and six-year-olds found that the number of 'conversational turns' the children heard was much more predictive of a child's language development, regardless of parental income or education.[22] Human adults communicate with babies by copying and repeating their gestures and babble (the so-called motherese used universally by parents of very young babies),[23] and this apparently inconsequential phase of oral grooming may well be a critical stage in human language development. It has a conversational rhythm with turn taking: the baby's verbalizations are repeated by the mother, with the same musical tone and pitch, back and forth in sequence. Babies become accomplished turn takers by around three months, taking just 600 milliseconds to respond.

Conversational turn taking is older than language: some primate and bird species also do it. Gibbons call in turns, for example, whereas great apes don't do it vocally but do it gesturally. Turn-taking species are all highly social and mostly pair-bonders, so they are invested in the other, and that investment is meted out in working out how the other is and what they are interested in. Turn-taking behaviours are used to smooth everything from mating to cooperative activities. In humans, it also enhances the collaborative nature of conversations that build on each other. Most of us obey the unspoken rules of turn taking, which are common to all languages, and it is very rare to explicitly come out and say that someone's hogging the conversation, unless we're addressing children. Instead, we use devices to rebalance an unbalanced conversation, such as interruption or making people laugh.

The sheer speed at which turn taking occurs in an average conversation means our mouths are jumping in before our brains have a chance to respond to what has just been said. On average, the response rate between speakers during a conversation is just 200 milliseconds, which is the fastest possible human response time, equivalent to the blink of an eye. But it takes at least 600 milliseconds for the signal to go from our ears to our brain, to understand what has been said, prepare a response and then send this to our mouths. The whole process of real-time conversational flow relies on our brain's sophisticated prediction system. We can only respond as rapidly as 200 milliseconds by predicting what someone's going to say and preparing to say something at the same time. So while one person is doing the speaking – most turns last just two to three seconds – the other has to decide what they're about to say in order to respond in time. Neuroscientists are still struggling to understand how we can do these two things at the same time, because a large part of the brain is involved both in speaking and listening. Nevertheless, we do conversational turn taking around 1,500 times a day.

Navigating our social world meant honing our prediction system to probe not just the physical world but also the inscrutable realm of other people's minds. Language may well have evolved because it is an unparalleled mechanism for allowing us to predict people in a larger society. It doesn't replace our other sensory input, such as gaze and body language cues that can overrule the words we are being told, but conversations between people build trust and alliances, spread reputations and foster good feelings between people. Turn taking is key to this.

During conversation, our prediction system uses grammatical clues (for example, 'if' is often followed by 'then'); facial expressions; pitch, tone, volume; and gestures (returning hands to the lap) to help us know when to cut in. Putting the key part of the sentence nearer the beginning enables early interruption, because the listener can be more confident that they know where the conversation is headed and what is about to be said. During this baton-passing process, a tipping point will emerge – the upswing of the baton passer's arm – and the listener begins thinking of a response, waiting for a place to interject. There will be a break point by the speaker of up to about 500 milliseconds at the zenith, and after that, if the listener hasn't taken the baton, the speaker will realize there's a problem. For example, if the speaker ends with, 'Do you want to go for a coffee?' and there is no response within 500 milliseconds (although Nordic people, for instance, have slightly slower response times), then the speaker might add to or qualify the question, to help the conversation continue: 'We could always go later in the week?' There is usually a longer gap for No responses than Yes responses. But we're predisposed to give positive responses as part of our evolved adaptation for cooperation, so it really is harder to say No. Imaging studies show our brains recoil from the word No.

Learning language is extremely complex and yet babies are very good at it. By the age of five, most children can speak fluently and command a vocabulary of thousands of words and unthinkingly observe the rules of their native tongue. We speak proficiently without needing to be taught the rules, the origin of our words, or to talk, and it's universal: children born profoundly deaf will spontaneously develop a sign language with complex grammatical rules, as rich as spoken language and which recruits the same neural pathways. Just as nobody invented the eyeball, our language is the result of the purposeless, mindless competence of cultural evolution responding to selection pressures, such as ease of pronunciation, learnability[24] and environmental factors.

This extraordinary, flexible communication system sprouts from the concept of relationships between things. So, at its simplest, if A=B and A=C, then B=C. This seems obvious, but it's actually a very sophisticated idea, which we must learn. There are nine of these relational categories, including opposition (up versus down), equivalence (a picture of a horse is the same as a horse) and comparison (an elephant is bigger than a mouse). All can be then generalized to apply to other situations. Having learned how the relational 'bigger' applies, for example, it's easy to identify the bigger object in other pairs, and generalize the rule to novel situations. This seemingly unremarkable ability, which children effortlessly acquire from the age of 16 months, is at the heart of language cognition, because relational categories allow us to transfer meaning between different items, so the word 'ball' can come to represent the object even if it doesn't sound like it and no ball is present. Eventually we are able to discuss abstract ideas, such as is it better to play football or to watch it? This is a uniquely human skill. Many other species can learn the basic rule of a relational category, but none can generalize its application. Even chimpanzees that have received extensive language training can't get it.

Once the 'rules' of word combination and relationships are learned, we can combine these symbols in novel ways, so language can evolve in a similar way to biological evolution, with words being analogous to genes. The extraordinary diversity and complexity of human language is the result.

Let me tell you a story:

> girl fruit pick   turn mammoth see
> girl run   tree reach   climb mammoth tree shake
> girl yell yell   father run   spear throw
> mammoth roar   fall
> father stone take meat cut   girl give
> girl eat   finish sleep

This story, by historical linguist Guy Deutscher, doesn't rely on any rules peculiar to English (in fact it violates them) or to the grammar of any other particular language, and yet you probably understood it easily enough. In fact, the story can be understood in any language. Deutscher used a few natural principles, which are rooted in the deepest level of our cognition, to compose it: grouping words together if the things they represent are closer together ('girl' and 'fruit'); ordering words according to the order events occur; and using the most common 'subject-object-verb' sequence (studies show we think first of the subject, then the object, then the action – only about 10 per cent of languages put the verb before the subject). So 'girl fruit pick' is easier to understand than 'fruit girl pick' or 'pick fruit girl' even though none of these follows English grammar rules of subject-verb-object.

It's easy to imagine how these simple organizational rules would have been used by pre-verbal humans to tell a story using gestures. By using relational categories, we would no longer need to tell the story in the place where the events actually occurred and with all the characters present: we could instead represent the elements. We didn't need formal grammar, either. Once we developed just a little shared vocabulary (24 words, here), we could tell this story and be understood. Consider that currently, around 25 per cent of our speech is made up of just 25 words. More than two-thirds of the world's languages use similar sounds for common words.[25]

From here, cultural evolution steadily increased the complexity of our proto-language, building vocabulary and adding rules to avoid confusion and make comprehension clearer. A recent study,[26] using artificial intelligence agents that transmitted random sentence strings between each other like human speakers, found that they picked up on any grammatical structure and generalized it, tending to produce more structure in their output than they received. After many generations, the agents' language produced the kind of structure that occurs in natural human language, emerging simply through its repeated learning and transmission.

Some grammatical innovations only appeared with the invention of writing, in the past 5,000 years, such as the conjunctions 'before', 'after' and 'because of', which allow for longer, more complicated sentence structures. Without these tools, the earliest versions of Sumerian and its contemporaries are repetitive and dull to read; with the innovation, clauses can be combined without losing the reader. Nevertheless, there are living languages without subordination tools, including some Australian and Arctic tongues. As with all products of cumulative cultural evolution, the fancier versions are created by the largest, most connected societies. Languages with more speakers tend to have more sounds, bigger vocabularies and they also diversify faster than those with fewer speakers.[27]

We can spot grammaticalization as it occurs. Nouns and verbs evolve a new use as adjectives and adverbs, and in time, they may lose their original uses and keep only their new meanings: the temperature 'rockets' up. Meanings continually evolve and change according to the social context they are used in. 'Nice', from the Latin for ignorant, used to be an insult and meant stupid in the thirteenth century. It went through many changes up until the eighteenth century with meanings like wanton, extravagant, elegant, strange, modest, thin, and shy or coy. Now it means pleasing or kind. Context is everything, however, and in some circles 'nice' is now a euphemism for dull. Metaphors make language sing and play a key role in the most mundane communication – abstract concepts could never emerge in a species in which everyone insisted that signals be literally true.

Just as genes are passed between communities, so too words and languages. We are flexible enough to have invented languages, such as Esperanto and sign language for the deaf; and, very occasionally, we repurpose old or extinct ones, as with Hebrew, a language used only for liturgy that was resurrected as the national language of Israel. These creations and reinventions are rare, whilst the steady alteration of languages happens all the time. For just as genes and organisms undergo natural selection, so grammatically irregular words are subject to a powerful pressure to 'regularize', which is why English is losing many irregular verbs. The proto-European 'drove', for instance, will evolve to the Germanic 'drived'.

Globally, it is young women who lead language change and innovation; male speakers lag behind by as much as a generation in some cases. Some of this is to do with societal sexism: women are not often in high-ranking positions that demand correct pronunciation. Young women are extremely sociable, which means that what they say gets heard by others; and men, when trying to attract women, talk using women's innovations. One example is vocal fry, or the 'creaky voice', which means speaking while constricting the larynx – an affectation used by Mae West in the 1930s, and revived by celebrities like Kim Kardashian. Social values, such as sexiness, become attached to certain linguistic features, and many people adopted the Kardashian version. Using 'like' as a conversation filler, or uptalk (where the sentence rises in pitch), are other changes made by young women that spread through society.

New pidgin dialects regularly evolve from word and grammar hybrids of pre-existing languages. Kiezdeutsch originated in Turkish migrant communities in Germany, but has now become a common way of speaking for young people who otherwise speak perfect German, including those with no Turkish origins. Like British teens talking 'Jafaican' – a mélange of Jamaican patois, Los Angeles rap-speak and South London slang (perfectly satirized by the comedy character Ali G) – Kiezdeutsch is strongly tied to identity and how the speakers see themselves in society. If elements of that language community appear glamorous or cool, then teenagers will adopt the dialect regardless of their ethnic or social background.

Meanwhile, the variety of accents in Britain, which in the fourteenth century rendered someone from Kent incomprehensible to someone from Norfolk, is diminishing. More people now sound as though they come from the southeast, perhaps because that accent is associated with affluence, the type of linguistic prejudice that George Bernard Shaw depicted in *Pygmalion*. But we all modify our language and our accents according to the people we are talking to or the situation, such as when writing a letter. Calculated or unconscious, these are attempts to appeal to the social group you are talking to. Expensively educated politicians give speeches in the 'Estuary English' of poorer classes – a reversal of Eliza Doolittle's attempts to talk herself up a class. Even the Queen is not immune, losing poshness over the decades: she no longer pronounces 'very' as 'veddy' or 'poor' as 'pooer'. And if the Queen can't be relied upon to speak the Queen's English, then who can?

Language is entirely interwoven with identity and cultural belonging. Small children copy native speakers of their mother tongue before being swayed by other cues, such as ethnicity. One of the reasons young women innovate with language is to gain strength by forming a bonded group of similar speakers, a clique that provides social support. Hearing someone use the same accent or language innovation as you do gives you confidence that you come from the same place, you share a social kinship, and that you are both likely to support certain cultural values and defend certain interests. Language is a strong signifier of group belonging, distinguishing different societies.[28]

Nowhere is this starker than New Guinea, the most linguistically diverse area on Earth, home to more than 800 different tongues. Topographic barriers between communities, such as mountains, swamps and rivers, have helped languages change in isolation, so now there are more than 1,000 different words for water on the island. Islanders also use language as a strong tribal identifier. One village collectively decided to change their word for 'no' from 'bia' to 'bune', in order to distinguish their speech from that of a neighbouring village. Another community on the island deliberately swapped all the masculine and feminine gender agreements in its language to be the opposite of its neighbours.[29]

The same process is occurring globally. There are currently more than 7,000 different tongues across the world, more languages for a single species of mammal than there are mammal species. Linguists have constructed language trees that trace our many twiglet tongues back in time to their branches of commonality, like Indo-European, which birthed languages from English to Sanskrit (but not Basque), and geneticists, archaeologists and paleontologists are using this information to trace the spread and diversification of humanity over time.

Once we can speak, we don't stop at one language. Most people on Earth are at least bilingual, and each language that a person speaks then changes their brain, personality and behaviour in subtle ways. Our cultural evolution of languages changes our biology.

'We are different people when we use different languages. Language has power over us. Our humour changes. Our body language changes. I like to use Turkish to write about sadness and English when writing satire,' explains author Elif Shafak.[30]

Language shapes the way we think. English speakers are better than Japanese speakers at remembering who or what caused an accident, such as breaking a vase. That's because in English we say 'Jimmy broke the vase,' whereas in Japanese the agent of causality is rarely used, and they will say 'the vase broke'. The structures that exist in our language profoundly shape how we construct reality – and it turns out that reality, and our human nature, differs dramatically depending on the language we speak. Our brains change and our cognition is rewired according to the cultural input we receive and respond to.

Take the evolution of colour terms. Societies generally start by defining lightness and darkness, and after black and white, the next named colour term is always red (presumably because it is the colour of blood). Red in English used to also include brown, purple, pink, orange, yellow.[31] Then the next colour terms are usually yellow or green. A lot of societies never get onto mentally registering blue, and learn for the first time that blue can be a colour category through learning English; many languages have borrowed words for blue. German speakers, who have many different words for hues of blue, are better able to distinguish between blues than are English speakers or the Himba people of Namibia, who don't have a word for blue and find it difficult to distinguish between green and blue. However, the Himba have more terms for light and dark tones, and Himba children distinguish between shades very easily – much better than Europeans.

In other words, our cultural invention of language influences our cognition to the extent that how (and whether) we have learned to verbalize the sensory input our brains receive (wavelengths of light) actually determines whether we consciously experience it.[32] We generate a word for a colour when we have two things that are identical except for colour. Industrialized societies have more identical objects to describe and to choose between than hunter-gatherers – we need to be able to distinguish the blue car from the green one, and so have developed a more expansive colour vocabulary. In a natural environment, choosing something of a different colour generally means choosing something with altered properties, which could equally be used as a descriptor. Some societies, such as the Jahai of the Malay Peninsula, don't have a large colour vocabulary but have a large odour vocabulary, and such people show much greater sophistication in differentiating between odours.

So, concepts that we consider universally human, like colour perception, interpretation of facial expression, notions of time or directions, are culturally learned through language and turn out to be surprisingly nuanced. Naming things opens a mental door into new cognition – new ways of understanding the world. Children who speak Hebrew, a strongly gendered language, know their own gender a year earlier than speakers of non-gendered Finnish.

Languages differ a lot in how they describe directions, too. In English we frequently use left and right: your left leg, for instance. But in around a third of languages, left and right aren't used. In the Guugu

Yimithirr language of Far North Queensland, which is the source of the word 'kangaroo', people describe positions and directions in terms of north, south, east and west: 'The boy standing north of Mary is my brother.' Every exchange needs to report direction,[33] so speakers must stay mentally orientated at all times just to speak in grammatical sentences. This requires cognitive changes in how we organize our language and spatial awareness. If you're going to tell a story, you have to remember if a person approached you from the west or east in order to tell that story properly, because in some of those languages any verb of motion will have to include the direction. It's an entirely different conceptual framework, which speakers of non-directional languages don't have, but which we can nevertheless pick up.

American anthropologist Lera Boroditsky did just that:

> I spent my first month in that community feeling pretty stupid because it's such a basic skill that everyone was treating me with a lot of pity. After about a week, I was walking along and . . . I noticed a little window appearing in my mind like you might get in a video game, and I was this little red dot you might get from a birds eye view map, and as I turned, the little window oriented itself relative to the landscape. Automatically, I thought, wow that makes it a lot easier. And then I rather sheepishly told someone about this . . . And they looked at me and said, well of course, how else would you do it? It took the social pressure of trying to appear normal in the linguistic community to drive my brain to create the solution that works so well.[34]

More than a century ago, it was established that our capacity to use language is located in the left hemisphere of the brain, specifically in two areas: Broca's area (associated with speech production and articulation) and Wernicke's area (associated with comprehension). Damage to either of these can lead to language and speech problems or aphasia. In the past decade, however, neurologists have discovered it's not that simple: language is not restricted to two areas of the brain or even just to one side, and the brain itself can grow when we learn new languages. Recent findings show that words are associated with different regions of the brain according to their subject or meaning. Work by neurologists[35] suggests that words of the same meaning in different languages cluster together in the same brain region.

Bilingual people seem to have different neural pathways for their two languages, and both are active when either language is used. As a result, bilinguals are continuously, subconsciously suppressing one of their languages in order to focus and process the relevant one. The first evidence for this came out of an experiment in 1999, in which English-Russian bilinguals were asked to manipulate objects on a table. In Russian, they were told to 'put the stamp below the cross'. But the Russian word for stamp is 'marka', which sounds similar to 'marker', and eye tracking revealed that the bilinguals looked back and forth between the marker pen and the stamp on the table before selecting the stamp. And it seems the different neural patterns of a language are imprinted in our brains forever, even if we don't speak it after we've learned it. Scans of Canadian children who had been adopted from China as pre-verbal babies showed neural recognition of Chinese vowels years later, even though they didn't speak a word of Chinese.

Multilingualism has been shown to have social, psychological and lifestyle advantages, as well as a number of mental health benefits.[36] Our brains seem to have evolved for multilingualism, which may well have been the norm in our deep past. Modern hunter-gatherers are universally multilingual. Many tribes have rules forbidding marriage with someone within their tribe or clan, so every single child's mother and father speak a different language. In Aboriginal Australia, where more than 130 indigenous languages are still spoken, multilingualism is part of the landscape. You will be walking and talking with someone, and then you might cross a small river and suddenly your companion will switch to another language, because people speak the language of the land. This is true elsewhere, too. Consider in Belgium: you take a train in Liège, the announcements are in French first. Then, you pass through Leuven, where the announcements will be in Dutch first, and then in Brussels it reverts back to French first.

Multilingualism has a surprising effect on the brain and the sense of self. Ask me in English what my favourite food is, and I will picture myself in London choosing from the options I enjoy there. But ask me in French, and I transport myself to Paris, where the options I'll choose from are different. So the same deeply personal question gets a different answer depending on the language in which you're asking me. This idea that you gain a new personality with every language you speak, that you act differently when speaking different languages,[37] is a profound one.

In one experiment, English and German speakers were shown videos of people moving, such as a woman walking towards her car. English speakers focus on the action and typically describe the scene as 'a woman is walking'. German speakers, on the other hand, have a more holistic worldview and will include the goal of the action: they might say (in German) 'a woman walks towards her car'. Part of this is due to the grammatical tool kit available. Unlike German, English has the -ing ending, the present participle, to describe actions that are ongoing. This makes English speakers much less likely than German speakers to assign a goal to an action when describing an ambiguous scene. For English-German bilinguals, however, whether they were action- or goal-focused depended on which country they were tested in. If the bilinguals were tested in Germany, they were goal-focused; in England, they were action-focused, no matter which language was used, showing how intertwined culture and language can be in determining a person's worldview.

In the 1960s, one of the pioneers of psycholinguistics, Susan Ervin-Tripp, asked Japanese-English bilingual women to finish sentences, and found great differences, depending on the language. For instance, 'When my wishes conflict with my family . . .' was completed in Japanese as 'it is a time of great unhappiness'; but in English, as 'I do what I want'. From this, Ervin-Tripp concluded that human thought takes place within language mindsets, and that bilinguals have different mindsets for each language – an extraordinary idea but one that has been borne out in subsequent studies, and many bilinguals say they feel like a different person when they speak their other language.

These different mindsets are continually in conflict, however, as bilingual brains sort out which language to use. This concerns a part of the brain called the anterior cingulate cortex (ACC), which is involved in executive control, allowing you to focus on one task to the exclusion of another. Brain-imaging studies[38] show that when a bilingual person is speaking in one language, their ACC is continually suppressing the urge to use words and grammar from their other language. In fact, it is possible to distinguish bilingual people from monolinguals simply by looking at scans of their brains: bilinguals have significantly more grey matter in their ACC, because they are using it so much more often. This enables them to perform better in a range of cognitive and social tasks, from verbal and non-verbal tests to how well they can read other people.[39] Bilingualism seems to keep us mentally fit and so perhaps it was selected for, both culturally and biologically – an idea supported by the ease with which we learn new languages and flip between them, and the pervasiveness of bilingualism throughout our history.

The key to our many languages is the innate desire within our socially driven brains to form strong cooperative groups in which the individual doesn't have to do battle with the world, but can rely on the tribe to. Talking creates and strengthens relationships between people whether they are kin or not, widening our social support networks. But the very success of our global social networks is driving language extinction – at an alarming rate of one every 14 days – because 80 per cent of the world's population can now converse using just 1 per cent of its languages.

We are now creating artificial intelligence to respond to our spoken instructions and even to talk to us. It has proved itself remarkably able. But language is more than simply encoded information, and robots are very primitive communicators. The reason for this lies in the subtle but profound difference between information and meaning. The information is embedded in the words and sentences, but the all-important meaning relies on the context – the cultural developing bath – of the speaker and the listener. That is why a sentence can be interpreted differently by different people, and why AI is not yet human. When Emily Dickinson describes hope as 'a thing with feathers that perches in the soul'; or John Donne declares, 'She is all states, and all princes, I'; or Robert Frost says that of two roads in a wood he 'took the one less travelled by, and that has made all the difference', they are easily understood by humans, but AI would be unlikely to process the information the same way. This is, incidentally, also applicable to genetic information, in which the message that is decoded depends on the chemical molecular 'context'.

Language gives us unparalleled ability to convey an infinity of ideas. We use it mainly to talk about ourselves, as we shall explore next.

8

Telling

You might expect someone who'd spent his childhood in a one-room school run by his mother and grandmother in Alabama to have rather limited horizons. After all, there's only so much you can learn from a handful of people in a 1970s sleepy agricultural town.

Little Jimmy Wales, though, had an escape portal: when he was three, his mother bought the World Book *encyclopedia from a travelling salesman. As soon as he could read, he was obsessed with it, compulsively going from one entry to the next, seduced by 'see also' references that led him along branching trails of information, ' "See also" – you could get lost in there,' he recalls.*[1]

Each year, World Book *sent out stickers with updates to the entries, and together, he and his mother pasted that year's advances in knowledge onto the pages. The experience seeded a bold idea.*

Four decades later, with the millions he'd made as a futures trader, Wales combined an interest in coding with his first passion and began commissioning experts to contribute articles for an online encyclopedia. It was slow, stultifying work, with a laborious system of exhaustive peer-review, until his newly employed philosophy graduate, Larry Sanger, suggested using a wiki so anyone could edit the pages. Instead of the usual top-down commissioning structure of publishing, wikis harness the creative potential of their many users, rapidly generating content.

Wikipedia launched in 2001. It currently has about 71,000 active contributors, working on more than 47 million articles in 299 languages, who update the site at a rate of ten edits per second. The English language version has more than 5.6 million articles, which is roughly 50 times Britannica's *count. Perhaps most remarkable is not the amount of content but the accuracy:* Britannica, *which employs experts including Nobel Prize winners to write its articles, is no more accurate in its scientific coverage than Wikipedia, which doesn't pay its writers or require they be qualified.*[2] *This shouldn't be surprising though, because Wikipedia is a microcosm of the same process humanity has been undertaking for hundreds of thousands of years: the accumulation, editing and updating of cultural information by society.*

Wikipedia makes cumulative cultural evolution visible. Language makes it possible. Language enables us to transmit detailed cultural information with very high fidelity, to many people at once, accelerating the evolution of diverse and complex technologies, societies, online articles, and more. Importantly, language also vastly improves teaching, and so its emergence would have been transformative in our ancestors' cultural evolution – indeed this may have been its main driver.

Language, like all communication, is inherently social. It helps us to strengthen and maintain social bonds in the most efficient, lowest-energy way, replacing one-on-one primate grooming sessions with small talk, flattery and gossip that can be carried out faster, while performing other tasks and across a whole group. Language helps glue our societies, boosting our individual survival and allowing us to cooperate in groups of millions rather than dozens. As our societies grow in size and complexity, we become increasingly reliant on reputational information to tell us who, of the many people that are unrelated to us, we should invest our energy, time and resources in.

Anyone can edit Wikipedia articles or generate new ones, which also means that anyone can insert errors, bias, or deliberate falsity. But for every incorrect statement, there are Wikipedians[3] ready and waiting to correct it or counter the bias – often within seconds. The success of the whole enterprise relies on reputation: facts are referenced with citations so users can judge the reliability of the source; editors themselves are ranked according to their experience; and, by extension, it also affects the personal reputation of those featured in Wikipedia articles, making them better known but risking unflattering information coming to light. The hundreds of millions of people who visit the site each month, outsourcing their individual memory and research costs to this collaborative effort, do so on the basis of its reputation.

By telling each other what is important and trustworthy, we use language to apply a powerful selection pressure on our cultural evolution, because it's through reputational information that we learn who to copy, what to copy, what to believe, and how to behave.

Why spend your free time writing articles for Wikipedia? Why help a bunch of strangers at all? The most convincing explanation is that altruism builds social cohesion, and as we've seen, humans are dependent on the social group for survival. The stronger our group is, the better it can compete for our interests against other groups, so the better our individual chances of survival. Cooperating – rather than competing – became so important for the survival of our genes that our default behaviour is to be fair and kind to each other, and we expend significant energy curating a reputation for positive social behaviour. Although moral rules vary between societies, as a species, across cultures, there are commonalities: we are respectful of each other's property, for example, and stealing from your group is a pretty universal no-no. Cumulative cultural evolution relies on social cooperation and altruism to produce complex and diverse societies, and the social tools to manage them cooperatively.

Biologists have in the past tried to pin our kind tendencies on the same evolutionary motivators that guide other animals: by being kind and helpful, we are helping our genes survive because we are directly or indirectly helping our kin. Altruistic animals, such as ants, are closely related, so altruism does help their own genes. And this is certainly true for plenty of human relationships and in very small societies. However, kinship alone doesn't explain the altruistic nature of most human societies, which are too big and diverse, with too many interactions with strangers, for us to be able to credit our selfish genes with our better natures.

Another suggestion for the evolutionary basis of our cooperative nature is that by doing someone a favour, they will return it – you scratch my back and I'll scratch yours. Reciprocal altruism makes sense for long-lasting relationships between individuals, but doesn't explain the vast array of anonymous altruism that we all practise every day, such as holding a door open for a stranger or bigger acts of charity, such as giving blood. We don't do this with the expectation that the strangers that receive our help will somehow find out who we are and compensate us. And yet our many acts of kindness are seen by others and copied. Our brains have evolved to be exquisitely attuned to social cues, with our so-called mirror neurons triggering in empathic response to another person's action or experience, and promoting imitative behaviour from the youngest babies. We are social copiers and get measurable pleasure from matching our behaviours and choices with those whom we like or admire.[4] This means that those who are generally liked – nice people – help make society nicer as more people emulate them.

In one study, when drivers let waiting cars out of a junction, those who had been let out were more likely to 'pay it forward' and let others out subsequently. Kindness contagion leads individuals to aspire to being 'better' people.[5] We wait in queues, hold doors for each other and cover our mouths when we cough; all these everyday acts of kindness cost us a little individually, but produce a helpful society in which we can expect doors not to be slammed in our faces. Over thousands of generations, this has domesticated us and made human societies generally cooperative, increasing group cohesion and hence the fitness of individuals. Cooperative people tend to be more successful, whereas selfish people have fewer children and earn less money.[6]

Some altruistic acts, though, make little sense from an evolutionary perspective. In March 2018, an Islamist gunman took several shoppers hostage in a French supermarket near Carcassonne. Police managed to persuade him to release all but one, a woman he threatened to kill unless his demands were met. In a supreme act of altruism, a police officer called Arnaud Beltrame requested to swap places with the hostage. The gunman shot Beltrame and he died. The woman he had replaced survived. Beltrame's altruism did not benefit his genes – the woman was unrelated to him. However, by his act of extraordinary kindness, he inspired good deeds by others, he strengthened the institution of policing, and his (posthumous) reputation was nationally recognized and celebrated, enhancing his family's social standing. Beltrame was acting as a police officer, a role created by society for public service. He was also a practising

Catholic, a religion that preaches sacrifice for others. Although this type of extreme altruism seems contrary to rules of genetic evolution, it makes perfect sense in terms of cultural evolution. Beltrame's altruistic act helped strengthen his group, generally improving the survival of those in it.

As we have evolved to be innately cooperative, it is generally less cognitively demanding – less costly in time and energy – to be nice, and that tends to be our default behaviour. This pays off because there are many occasions in which acting in self-interest is not actually in our interest – statistically, cooperating works out better. This is best explained using a classic thought experiment called the Prisoner's Dilemma. Two members of a criminal gang are imprisoned in separate cells with no means of communicating with each other. The prosecutors lack enough evidence to convict the pair so they are offered a bargain: testify against the other, or remain silent. If both betray the other, each of them serves two years in prison; if one betrays but the other stays silent, then the betrayed gets three years and the other walks free; if they both remain silent then each gets one year in prison. While it might seem rational to betray the other,[7] if they both acted in self-interest, then they'd both get two years – the combined worst sentence of four years. In fact, their best option is cooperative silence. And this is the case in enough real-world situations for the cooperative strategy to have evolved as our behavioural default.

Studies show, for instance, that when people are asked to rapidly donate to a collective pot in a so-called public-goods game, where the total will be redistributed equally among the players, they will automatically give generously. They do so even though this sort of social dilemma, like all cooperation, relies on trusting that the others in their group will also be nice. In a four-player public-goods game, if everybody in the group contributes all of their money, all the money gets doubled, redistributed four ways, and everyone doubles their money. Win-win! But even though everyone is better off collectively by contributing to a group project that no one could manage alone – in real life, this could be paying toward a hospital building, or digging a community irrigation ditch – there is a cost at the individual level. Financially, you make more money by being more selfish. From the perspective of an individual, for each dollar that you contribute, it gets doubled to two dollars and then split four ways – which means each person only gets 50 cents back for the dollar they contributed. It makes financial sense, then, to give the least possible to the pot and freeload off everyone else's generosity. If players are given time to consider their decision, they often overrule their automatic gut response to be nice and become less generous.

Whenever we help strangers we have to overcome the possibility that they may exploit us. We use stick and carrot strategies, as a society, to curb this problem. In the long term, people mostly benefit by cooperating with the group, even if they sometimes incur costs, so it is in their interest to stay. This gives society power over their behaviour: being allowed to stay in the group and benefit from it is contingent on an individual's reputation for behaving cooperatively. In the small-scale societies that our ancestors were living in, all their interactions were also with people that they were going to see again and interact with in the immediate future. That reputational threat helped keep in check the temptation to act aggressively or take advantage of other people's contributions.

Cooperation breeds more cooperation in a mutually beneficial cycle, but so can the opposite – we can learn not to be nice. Our innate desire to cooperate is shaped by society, and throughout our lives we learn to modify our helpfulness. Those in the experiment who played the quick-fire round were mostly generous and received generous dividends, reinforcing their generous outlook. Whereas those who considered their decisions were more selfish, resulting in a meagre group pot, reinforcing an idea that it doesn't pay to rely on the group. So, in a further experiment, the researchers gave some money to people who had played a few rounds of the game, and asked them how much they wanted to give to an anonymous stranger. This time, there was no incentive to give; they would be acting entirely charitably.

It turned out there were big differences in generosity. The people who had got used to cooperating went on to give twice as much money as those who had got used to being selfish. A brief experience of the benefits – or not – of cooperation had changed people's internal compass and behaviour, even with no institution in place to punish or reward them.[8] This reveals how plastic the human mind is, and how important our cultural developing bath is in shaping our behaviours, even if we are born with an innate tendency toward a particular behaviour.

The same research team, at Yale's Human Cooperation Lab, has also tested how people in different countries play the game, to see how the strength of social institutions – such as government, family, education, and legal systems – influences individual behaviour. In Kenya, where public sector corruption is high, players initially gave less generously to the stranger than players in the United States, which has less corruption. This suggests that people who can rely on relatively fair social institutions behave in a more public-spirited way; those whose institutions are less reliable are more protectionist. However, after playing just one round of the cooperation-promoting version of the public goods game, the Kenyans' generosity equalled the Americans'. And it cut both ways: Americans who were trained to be selfish gave a lot less. So, our cultural developing bath influences our co-operative behaviour but we are cognitively flexible enough to adapt quickly to other social environments.

Whatever our wider social environment, human groups are not homogeneous collections of people, but rather complex networks of individuals, and how thoroughly interconnected that network is affects how behaviours and information spread through it. In some networks, like a small isolated village, everyone is closely connected and a villager is likely to know everyone at a party; in a city, by contrast, people may be living more closely to more people, but the city dweller is less likely to know everyone at a party there. The different properties of these networks influence the group's behaviour generally and that of individuals within them, as is clear if you've visited cities and villages. Social psychologists are now probing these effects by manipulating the shapes of social networks and the positions of influential people in them. In one experiment, Nicholas Christakis's team at Yale's Human Nature Lab created temporary artificial societies with online players to see how they would interact and how kind they were to each other. Then he manipulated the network, changing the way people were connected. 'By engineering their interactions one way, I can make them really sweet to each other, work well together, and they are healthy and happy and they cooperate,' he said. 'Or you take the same people and connect them a different way and they're mean jerks to each other and they don't cooperate and they don't share information.'

In one experiment, he randomly assigned pairs of strangers to play the public-goods game with each other. In the beginning, he says, about two-thirds of people were cooperative. 'But some of the people they interact with will take advantage of them and, because their only option is either to be cooperative or to be a defector, they choose to defect because they're stuck with these people taking advantage of them.' By the end of the experiment, he said, 'everyone is a jerk to everyone else.' Christakis turned this around simply by giving each person a little bit of control over whom they were connected to after each round.

'They had to make two decisions: am I kind to my neighbour or not; and do I stick with this neighbour or not,' Christakis says. The only thing each player knew about their neighbours was whether they had cooperated or defected in the round before. From this, he was able to show that people cut ties to defectors and form ties to co-operators, and the network rewires itself into a cooperative prosocial structure instead of an uncooperative structure.[9] These experiments help reveal how cooperative societies have emerged out of generations of human interactions.

We police our societies through reputation, by punishing mean behaviour and by cutting social ties to uncooperative people. We also carry within us our own reputational police in the form of conscience. We are able to empathize with another person and act in a way that we would find kind or helpful if we were they. In a recent brain-scan experiment,[10] in which people were offered cash in exchange for painful (but harmless) electric shocks being applied to them or to a stranger, they experienced less pleasure when they got the cash at someone else's cost than when they got a smaller payout but suffered the pain themselves. Ill-gotten gains are not valued as much by our brains as honest earnings. During childhood, we develop the self-awareness that allows us to see ourselves as others see us, and modify our behaviour accordingly. A few highly intelligent, social animals can also achieve this so-called 'theory of mind' capability to some extent, although none is as well developed as ours, and we are not born with it.

In a classic experiment, a young child is shown a doll and two lidded boxes. An adult enters the room and hides the doll in the first box, then leaves the room. A second person enters, takes the doll out of the first box and hides it in the second one. The first adult returns to retrieve the doll, and the observing child is asked which box she will look in. Young children will point to the second box where the doll is hiding. Only once a child is around four years old[11] will she realize that her own knowledge of the room is different from the adult's, that she has a different perspective from another person. Once this happens, she acquires enormous social power to manipulate the minds of others, to tell stories to another person that may not be true and which can be deliberately skewed for her own advantage. Telling a lie is cognitively demanding: you must be able to conceive of an alternative reality and describe it, then keep that in your mind along with what really happened, and distinguish between both, and you also have to be able to understand the idea that your listener has a different version of reality, and hold in your head their version and what they know. It's exhausting. One theory holds that our big brains evolved out of an arms race to optimize this power – our Machiavellian intelligence. Primatologists have observed a strong correlation between an ape's propensity to deceive and the size of its brain.

In a socially dependent species there is an evolutionary advantage to being able to manipulate others to our advantage, and we grow up to become expert manipulators. This ability forms the basis of our jokes, our stories, our politics, and less benignly our crimes against each other. Nevertheless, on the whole, we are helpful, kind and feel morally obligated to be thoughtful of each other's needs. Trustworthiness and altruistic, kindhearted characteristics are highly valuable traits in our society, translating into real economic advantage.

We all benefit from a generally nicer society because in most social situations people's interests are at least partially aligned. As our ancestors' groups got bigger they needed to cooperate more with non-kin, who were less invested in their welfare, and social skills became increasingly important. Being able to manage a greater number of social relationships allows us to live more efficiently, cooperate over resources and find mates in a bigger gene pool, which enhances our reproductive success. It also increases our pool of cultural resources, which improves survival.

However, while larger groups have their benefits, they are also more stressful and competitive, which makes them more cognitively demanding social environments. The alliances that must be formed, guarded and nurtured; remembering everyone's position in the social strata and their reputations; and understanding who can be trusted take time and effort at the expense of self-care, hunting and other activities. It's no coincidence that the bulk of the dramatic increase in brain size over our evolution has been in the neocortical region, where social cognitive processing occurs, and via increased cortical folding, which enhances the connectivity required for language. Larger groups provided selection pressure for the evolution of language, and the evolution of language made larger groups possible in another evolutionary feedback loop.

In the 1990s, evolutionary anthropologist Robin Dunbar found a robust relationship between community size and neocortical size across primates – the so-called Dunbar's number.[12] The neocortical size of most apes limits their group size to around 30, with bigger-brained chimps managing a social circle of 50 to 60. Brain size more than tripled over human evolution, and our own neocortical size corresponds to a Dunbar's number of 150 for the number of people with whom we can maintain a meaningful relationship, involving trust and obligation.[13] This seems to tally, whether we look at population records for Domesday villages, modern hunter-gatherer societies, Christmas card send-outs, or Facebook relationships,[14] although there are signs that Internet communities may stretch it beyond 200 (we are now exposed to so many different faces, our brains can recognize around 5,000 people).[15]

For our primate cousins, grooming is extremely time consuming and grows very tricky in larger communities. Chatting was our species' answer to managing our large social life. Indeed, studies on chimps show that in novel situations where they are dependent on another individual to act, they simply amplify the same calls they use during grooming, indicating that grooming and 'chat' are to some extent interchangeable – primates living in the biggest groups have the largest vocal repertoire. For us, gossip does the job of grooming, and much of our small talk or banter is phatic, where the words themselves are less important than the collaboration being sought.

We talk about the weather to maintain the social bonds that enable us to cooperate in groups of unrelated people. The goal is to make your listener feel good in your company and to like you, but it's a learned ability and young children often struggle with it, giving literal answers to phatic questions such as 'How's it going?'

Through chitchat people find common ground. Such conversations build on commonality and generate good feelings and shared experiences, thereby condensing into a short time what would otherwise take many days of joint activity – chatting reduces the time and energy costs of generating vital social bonds. And we've evolved to love it: the reward center in our brain is activated when we share our opinions or information, making us feel good. Humans have long childhoods and longer life spans, and over the years individuals will often need help, so it pays to form relationships with reliable individuals outside of immediate family.

At least 60 per cent of our conversation is gossip about people not present, during which we discover – and create – their reputations. Reputation reverberates the consequences of our actions beyond their immediate occurrence, and is a socially created labour-saving device that allows us to sample other people before we buy into them. People are consistent: how they've behaved in the past tends to be a good guide to how they'll behave in the future.

Trade, for instance, requires considerable trust. If you are going to barter a sheaf of carefully crafted arrows for a leather cloak, you need to trust that the person you handed over the arrows to will make good on their part of the deal (once they've used your arrows to bring down the bison that's wearing the cloak). In small, tightly knit societies, this is relatively easy, but as groups grow, it becomes increasingly precarious. You have a reputation only if you are well connected and your networks are interconnected. Our extended families helped with this – as we are the only apes that recognize in-laws and their relatives by marriage as kin, this allows us to spread our networks. Our 150 Dunbar number can straddle several groups and, because we have language, and therefore names, we can forge connections through friends of friends, kin of friends, and so on, using their (and our) reputations to extend our networks ever further across different tribes and cultures. In this way, we can connect cooperatively as individuals even if the tribes or societies we each belong to are competing in animosity.

In a world where our survival and the success of our genes depends on our position in a complex human society, reputation is incredibly important. A good reputation delivers us an upgrade in all our dealings: we are more likely to be offered help and assistance when we need it, and our children are more likely to be looked after. A poor reputation, by contrast, can lead to that most terrible social punishment, ostracism, and ultimately death. But although we can help shape our own reputation, we cannot control it entirely; it can even survive our personal extinction. And it is difficult for someone to judge a person on their actions alone if they have heard stories that paint a persuasive picture, because our social learning is overwhelmingly based on copying others rather than innovating our own ideas and opinions. Experiments show that when people are playing trust games with strangers, even after several rounds, their opinion on the stranger's trustworthiness (based on experience) will be swayed by the opinion of a previous player (who had similar amount of playing time with the stranger).[16] So, if they could see how the stranger had played in the past, they would cooperate with him around 60 per cent of the time, but if this was augmented by a positive snippet of gossip, cooperation went up to 75 per cent. Negative gossip, however, meant cooperation dropped to 50 per cent, even if it contradicted the evidence the player could see with their own eyes and the gossiper was as inexperienced as they.

The pressure to be agreeable makes us wary of expressing an opinion at odds with the pack, and keen to be seen to affirm a popular member of our group. This can lead to progressively more extreme opinions, to social media 'pile-ons' in which someone with a previously clean reputation has it trashed over a small mishap, or to individuals developing a cultish following. In small societies, gossip can make or break an individual; in large societies the stakes are higher. The battle to control and command reputation has led from the absurd to the extreme: from Ramses II declaring every battle he fought an Egyptian triumph, to China banning news websites and media. Our innate reliance on social information transmitted via gossip makes it a powerful tool for people who wish to enact social change by tarnishing the reputations of individuals or groups. A timeless joke from the 1930s concerns a Jewish man merrily reading *Der Stürmer,* a Nazi propaganda rag. In answer to his friend's confusion, he says, 'If you read the Jewish papers, everything looks terribly black for us, but according to this, all is well! We control the banks, the country – we run the whole world!'

Cultural edicts warn against bearing false witness and badmouthing people – what Roland Barthes called 'murder by language'. However, gossip is an essential tool for policing our interdependent societies, bringing wrongdoers and selfish and antisocial people into line, and making sure everyone in the group is pulling their weight. The downside of gossip is that anyone can become a bully; the advantage is that anyone can spread gossip. You don't have to be tough enough to physically take someone on. In this way, gossip can rectify antisocial behaviour without violence.

When we are watched, we are more likely to behave ourselves.[17] Burglars have been known to turn their victims' family photos face-down – no one wants to be seen being bad; and in the same way, simply putting up a picture of a watchful pair of eyes reduces shoplifting.

The gods of the newer monotheistic religions are the ultimate judgemental eyes, spying on our every move and deciding whether our actions will damn us to hell or allow us to enter paradise. The Jewish, Christian and Muslim prayer books all reference divine surveillance of a judgemental god that can see into your heart and mind – most gods are much more interested in bad deeds than good ones. It is likely that our religions evolved through a social selection pressure to police our growing societies.[18] As we saw with the Homeric gods, the type of religion a society adopts seems to depend partly on the type of policing it needs. High gods, who take an active role in human affairs and morality, are more common in large societies with money and taxation that require large-scale cooperation between strangers. Indeed, beliefs in punitive, interventionist gods may have evolved as an adaptation to facilitate large-scale cooperative behaviour across geographically distant peoples.[19] Social anthropologists recently tested this theory using an online game in which individuals allocated money among themselves, local and geographically distant co-religionists and religious outgroups, including those practising Buddhism, Christianity, Hinduism, and also forms of animism and ancestor worship. They found that believers in morally punitive gods were more generous to distant co-religionists (and less influenced by other similarities, such as how geographically close they lived).[20] Moralizing gods help expand cooperative behaviour – and our drive to enhance a reputation with an all-seeing god may compensate for society's weaker reputational force as communities grew bigger. Religious people are perceived to be nicer, more cooperative people,[21] but although they are more likely to be trusting and trustworthy, they tend to limit these qualities to those who share their values.[22]

Our cultural use of the emotions of reputation – shame and guilt[23] – may also have evolved with our ancestors' move to bigger societies. Neither has been observed in apes, yet they are innate and universal in humans. Withdrawing someone's self-worth by shaming them has a powerful physical as well as psychological effect. The body responds to shame as it does to a physical wound, with a surge in the stress hormone cortisol and an inflammatory response, which can be damaging if prolonged.

Many societies retain shame as a primary influence over behaviour – Japan, for example, is a shame culture, in which the force of others' opinions is a stronger influence on behaviour than guilt. Whereas in guilt cultures, such as the United States, people rely on guilt avoidance and conscience more than shame. The relative importance of shame or guilt as the main morality driver seems to depend on how closely networked[24] the society is for gossips. In tight-knit societies with long-lasting social bonds and little anonymity, such as gossipy villages, people become judgemental and often attribute social differences to character flaws or virtues. Shame is an important method of social control in such groups and the best way to appease is to conform. However, in individualistic societies such as cities – where people are more private and less densely connected and individuals are reliant on many overlapping groups rather than a single group – the pattern of gossip may produce less judgemental attitudes and so shame is weaker and less effective. Instead, internal promptings of guilt may be more successful.

When people are devalued by others, they devalue themselves – our self-esteem depends on how others see us, numerous studies have shown. In other words, our self-esteem depends on our reputation, which itself becomes a driver of moral behaviour. Similarly, if our conscience is clear, our self-esteem grows, and others judge our high self-esteem as being associated with a good reputation, which drives high self-esteem, and so on. Our personal morality steers us toward actions that will make others think well of us, raising our self-esteem. This kind of introspection is cognitively demanding, but enables us to manipulate others in social situations.

A few years ago a satirical British documentary[25] mocked society's judgemental attitudes to HIV transmission – haemophiliacs who had acquired the autoimmune disease through a blood transfusion were described as having 'good AIDS', whereas those who had caught it through sex or injecting drugs had 'bad AIDS'. Like the best satire, this tapped into a very real, if ridiculous, value system that has consequences. One study[26] found that gay men with HIV showed higher viral loads and faster declines in immune cells, and died on average two years earlier, if they were very sensitive to social rejection and felt shame about their illness. These emotions are painful but evolved for a reason – they show empathy, which plays a key role in our efficient social learning and cooperation, and show that we value the opinions of the group we belong to. Our compliance with the social values of our group is the price we pay for the benefits of belonging. Social ostracism is death, and a person who doesn't show shame or embarrassment doesn't care about social acceptance, making them dangerous and untrustworthy to have around.

Parasitizing others' experiences is by far the best way to acquire information, as we've seen. We don't need to sample all the options when deciding which restaurant to visit; we can copy the majority and use the popularity (reputation) of a busy rather than empty restaurant to guide us. Occasionally, this compulsion to copy others can be disastrous, such as when it causes a stock market crash, but more often than not the side effects are harmless fads and fashions. In general, social information (gossip) is a useful guide to reliable cultural information.

Reputation tells us who to copy. After all, if we copy the wrong person, we could sicken or become malnourished. And this poor rendering may then be passed down to the next generation who copy us. Instead of our technologies and cultural achievements evolving to improve in design, complexity, and diversity over the generations, they could become worse, with skills lost and techniques becoming rougher. Reputation applies a selective pressure that makes cultural evolution more efficient – it filters out noise and amplifies more reliable options.

All social animals have to make decisions about who to copy, but we humans are far better at it, and we tend to follow the same patterns globally. In infancy and early childhood, we learn initially from our parents and then from older siblings. We preferentially copy people of the same sex, language and culture,[27] and as we grow into adolescents, our peers become increasingly important – this modulates what we have learned from our elders, keeping our knowledge relevant for the time and societal changes. However, we don't necessarily choose who to copy based on their competency at the task. So, for instance, one study looking at fruit choice among schoolchildren showed that they copied the fruit choice of older rather than younger children. But after all the children had been tasked with a puzzle to solve, they switched their choice of who to copy in favour of those children who happened to be good at solving the puzzle, even if they were younger. This reveals much about the transference of prestige.

Prestige is a particular form of status that only humans recognize. Most animals observe the advantages of dominance, such as being the most powerful or aggressive or virile individual, and these are also important for humans: fearsome warriors are feted universally. Prestige is almost the opposite. Prestigious individuals are ones worthy of learning from – they are experts, older people. And if someone has prestige in one field, they become high-status individuals and their influence won't be limited to their field; we are likely to copy all their decisions. Indeed, prestige may have evolved as a way to enhance the benefits of cultural transmission.[28] Success in one area of life affords someone the status of general opinion leader. We want to learn from successful individuals, or even simply be associated with them in some way, so that their reputation rubs off on us. That's why a golfing hero can sell you a watch.

This may have its roots in the complexity of our cultural techniques. For example, being a good hunter requires several skills – being able to run fast, to track, to use weapons accurately, to cooperate well in the group, to bring down a beast. Learners can recognize a good hunter but can't be sure which skill makes him so good, so the best thing to do is to emulate the person.[29] However, copying someone's behaviour because of their reputation in another field can lead us into danger, such as when a celebrity commits suicide. Copycat suicides have occurred in several countries, often by people with no prior hint of depression, who copy the exact method and other details of the initial tragedy.

Prestigious individuals hold great power: they can reshape a social network, making it prosocial or uncooperative, tolerant or insular. When Princess Diana embraced AIDS patients, she had far greater influence on society's attitudes and understanding of HIV transmission factors than a decade's worth of lectures by virologists. Similarly, when a politician fails to condemn racial hatred – or even condones it – it emboldens others to copy him and, particularly if he is the national president, resets society's moral codes, the cultural developing bath, for a generation.

Because our self-worth is based on how others see us, prestigious individuals often buy into their own reputation, believing not only that they are superior in a particular field, but that their superiority extends to all other areas too.[30] Many celebrities limit their social circle to fellow celebrities and uncritical admirers, and consequently feel confident about proclaiming unwarranted expertise, such as actors that endorse dubious medical treatments.

Different cultures bestow prestige differently. Among hunter-gatherers, the best hunters' opinions on all matters carry more weight. It makes sense to copy older people, not just because they have had longer to accrue information, but because merely getting old was an accomplishment in the societies of our evolutionary past. By the time ancient hunter-gatherers had reached 65, natural selection had already filtered out many of their cohort, so their practices have more value, as evolutionary anthropologist Joseph Henrich neatly explains using chili peppers. Imagine a community with 100 people between the ages of 20 and 30, of which 40 routinely prepare meat dishes using chili peppers. Using chili, by virtue of its antimicrobial properties, reduces a person's chances of dying from food-borne pathogens. If eating chili year after year increases a person's chances of living past 65 from 10 per cent to 20 per cent, then the majority of this group, 57 per cent, will be chili eaters by the time they reach 65. If learners preferentially copy older rather than younger people preparing meat, they have a greater chance of acquiring this survival-enhancing trait, which, within a few generations of cultural evolution, will make chili inclusion a usual step in meat preparation by the group. 'Age-based cultural learning can thus amplify the action of natural selection, as it creates a differential mortality,' Henrich explains.

In Western society, perhaps because we are routinely living longer and because of the pace of recent technological change, age has lost prestige. Rapid cultural change makes social learning less reliable because you risk copying someone with outdated information. Nevertheless, in certain fields, such as creative endeavours that require mastery of particular skills, age is still revered. It takes minutes for a master potter to throw a perfect pot, but a lifetime to gain the skill.

It is no surprise that across all cultures, the most prestigious in society are those who possess the best knowledge and are the most generous in sharing it: our teachers. Teaching is all about communication, and the tools we invented for this also strengthened our cooperative societies, binding us in joint purpose through shared stories. Our group identity is woven into the words we speak, but this also makes language a vital tool for bridging cultural divides, because conversing with another group in their tongue breaks down distrust and enables access to their society. Our success as a species rests both on the competition between our prosocial groups and the necessity of cross-cultural exchange, as we shall discover.

BEAUTY

We are made human by the contemplation of beauty. We seek meaning in our lives and beauty gives us purpose, even immortality, through its expression. Beauty is subjective, invented, but it motivates us to act in ways that affect our evolution. Our greatest collaborations have been driven by beauty; it has enabled us to become global players. Beauty built our human world, or, as Ralph Waldo Emerson put it: 'The world exists to the soul to satisfy the desire of beauty.'

9

Belonging

I have an old wardrobe in the corner of my bedroom onto which I have screwed two ceramic knobs, and from these I hang my necklaces. Strings of polished stones, shells, and metal beads gleam and sparkle in the moving sun. Chains of interlocking silver hoops glint and liquify the light. But the translucent beads – glass, plastic, cut stones – have a magic all their own. How they shatter the sun into a thousand colours, throwing it out from their tiny hearts, transforming the dull wardrobe to a dancing waterfall of shimmering iridescence.

My children are transfixed. They lift the necklaces reverentially, letting the chains pour through their hands like articulated ribbons. They examine the beads one by one, wondering at their differences, marvelling at their clarity in the light. They plead to wear them, just to try one on for a moment, please Mummy. Their glee as I place one over a head and fasten a clasp around another's neck. The way they stand a little taller, shoulders back, tiptoes to the mirror.

They are cheap, my necklaces. I have just one that could be described as valuable to those who buy and sell such things. It was given to me by my grandmother and, because she has since died, it is an heirloom. The gold chain is strung with a beautiful black pearl, set in a gold socket. It is quite unlike anything I would choose for myself, and yet I treasure it for its embodied meaning. It was a gift from someone very dear to me, and before that, a gift from my beloved grandfather to her. It carries memories, and it is also strange and flawless. I wear it occasionally to intimidating events, where I take comfort from its solid weight around my neck and the timeless-ness of its beauty. Its provenance feels symbolic, a metaphor for dealing with life's challenges: pearls are the oyster's response to discovering grit in their shells. They are hard to find, rare at this size, and often involve dangerous oceanic dives in exotic locations. My necklace is made of multiple parts, of traded different materials, involving skilled people in different locations. It took someone with imagination to visualize how its constituents might be brought together and assembled into a pleasing and valuable whole.

My other necklaces are strung with glass, plastic, wooden or ceramic beads, shells, buttons and other cheap materials, but they have worth for me. They are pretty to look at and they prettify me when I wear them. Some are souvenirs of a time and place, and mentally transport me back to that experience. I have a brightly coloured string of plastic beads that were thrown to me during a hot, damp Mardi Gras in New Orleans, many years ago. They remind me of that heady trip, of my first solo journey across the United States in my early twenties. Through the beads, I recall the clamour of people on the streets, the dancing and the music, the slightly dangerous edge to it all. The necklaces are part of a centuries-old French-inherited tradition there, thrown by men at women in exchange for dances, beer, or a flash of nudity. I was thrown the necklace by a beautiful young man dancing topless on a balcony. I caught it and he called down to me: 'Show me your tits!' Appalled, I ran down the street and into a one-room bar, crowded with rhythmically writhing bodies moving in concert to the rhapsodic energy of a three-piece band. I stood there for a while, captivated, clutching my beads in the steamy air, becoming an adult while the music washed over me. When I hold the cheap plastic necklace now, I am holding my only link to the person I was in that other time and place.

My necklaces are decorative objects whose precious symbolism is known only to me or someone very close to me, but this is unusual. Jewellery often has another, overtly symbolic function, showing the wearer to be wealthy or from a certain tribe – wearing a cross marks someone as belonging to the Christian faith, a ring on the fourth finger shows its wearer is married. Even in my case, the type of jewellery I wear sends subtle messages about my lifestyle, age, background, social class, gender, and so on.

Beautiful things give us pause and invite us to contemplate them.

We have an emotional – biological – reaction to beauty and human culture has exploited and nurtured this, enabling us to ascribe meaning and value to decorative expression. We use this subjective meaning as a tool to create cohesive tribal societies organized through culturally agreed symbolism, norms and rituals. These invented norms evolve under social and environmental pressures, and powerfully affect our biology and genes. They reshape us and our societies.

We use beauty to signify belonging in large communities of genetically unrelated people. Beauty allows us to create artificial phenotypes that then influence our evolution.

We are strongly receptive to beauty. We seek it everywhere – in the faces of others, in the perfect symmetry of a flower, in sweet birdsong, in our created compositions – and we derive pleasure from recognizing it. Beauty has the power to console us; it can bring meaning and purpose to our lives; it can increase our powers of empathy and generate a sense of community. Beauty begets beauty, so that a neighbourhood that is decorated or planted with flowers impels people to continue enhancing the space.[1] We can appreciate beauty where we find it, but we are also motivated to create our own expressions of beauty through art, music, architecture, literature and dance, and in our material culture. Indeed, most of what we do or make has been coopted by our drive for beauty – our activities interwoven with ritual, our made objects designed for aesthetic meaning. Whenever we eat, we observe table manners, we talk at a socially acceptable volume while avoiding 'ugly' words, and we dress our bodies before appearing in front of others.

Humans spend enormous amounts of time and effort pursuing beauty, to the extent that we may die for our art. In 2015, Syrian archaeologist Khaled al-Asaad was beheaded by militants for refusing to reveal the location of Palmyra's ancient artworks. For the 81-year-old, the beautiful stone statues and columns of the 2,000-year-old temple complex were worth defending with his life.

Beauty is a powerful social tool and yet it doesn't exist in and of itself, and it is subjective. Probably, our invention of beauty has its roots in the biology of sexual selection. Peacocks are among many birds that use extravagant displays to advertise their biological fitness – an animal that can afford to waste its energy on such frippery as iridescent tail-feather 'eyes' clearly has an abundance to waste. For this reason, peahens have evolved to prefer males with the most extravagant tails. Humans are different because both sexes get to select their sexual partners, and so, as you might predict, we seek beauty standards in both male and female faces that, like peacock feathers, are markers of health and hard to fake. These include a high degree of facial symmetry and a flawless complexion.[2] Other primates also use faces to select for genetic quality: rhesus macaques, like us, prefer mates with symmetrical faces, for instance.

Studies show that humans find composite faces, made by an algorithm to an average of the population's faces, more attractive than the average individual face.[3] The evolutionary roots for this preference may be that good genetic mixing generally offers greater resilience and adaptability to the environment. In surveys, people generally find 'mixed-race'[4] individuals attractive and inbred families less so.[5] Fertility signals are also attractive traits: in males, these are indicative of higher testosterone, and in females, of higher oestrogen.

Our sense of beauty is therefore based on more than an aesthetic whim. People prefer young, healthy, fertile mates with no sign of disease – this combination preferentially activates our desire to mate, and we describe those individuals as more beautiful. Those who were best at identifying the healthiest, most fertile partners would have passed down their genes in greater numbers. So our sense of beauty and our actual beauty has improved over thousands of years.

However, many of our beauty preferences are subjective and have no basis in our objective biological fitness. Indeed, they seem to occur on a whim and be led by fashion. Again, there are intriguing parallels in the animal world. In the 1980s, evolutionary biologist Nancy Burley, who was working with zebra finches, started fitting small coloured bands on her birds to help identify different batches as they came into the lab. To her surprise, she discovered[6] that birds wearing bands of certain colours were more successful at finding mates, and put more effort into parenting their offspring. Females preferred red-banded males, while males preferred black- and pink-banded females. The zebra finches had 'evolved' a new set of sexual ornaments in the lab, fast enough for Burley to watch it happening. These bands were meaningless as fitness signals, and imply a certain randomness about which traits animals evolve to find beautiful – perhaps certain traits or colours are hard-wired into their brains, predisposing them to prefer certain novel mutations when they pop up. It could be that much of the diversity and beauty that we see in the natural world is down to animals' own appreciation for beauty.

Such seemingly random preferences seem to have shaped human appearance too. For hundreds of thousands of years, humans lived in small populations of different tribes, where cultural and genetic differences were able to accumulate. Over millennia, these have produced noticeable differences in how people look from Sri Lanka to Sweden. In small populations, the rate at which traits accumulate changes – some gene types may be lost altogether because there aren't enough people in the group carrying them, whereas others may become unusually common because the group happens to contain many carriers. It is likely our varied hair colour and eye shapes arose in small populations, and persisted because the people there liked those fashions and chose their sexual partners on that basis.

Take the thicker hair, more numerous sweat glands, distinctive teeth and smaller breasts of East Asian populations, all of which link back to a mutation in the EDAR gene that occurred some 35,000 years ago. Experts divide over whether that gene spread so rapidly because it emerged during a hot climate, when extra sweat glands would have been useful, or whether people just found those features more appealing. Fair skin[7] and blue eyes may also have once stood out as exotic and attractive, and such people may well have found it easier to get sexual partners, helping the traits to rapidly spread in northern Europe. Over the past 2,000 years, Britons have become taller, blonder and more likely to have blue eyes.

Attractive faces activate separate parts of our brain's visual cortex that are especially tuned to processing faces and objects. At the same time, the brain's reward and pleasure centres are activated, even when we're not thinking about beauty. There is also a moral component to our appreciation of beauty, with overlapping neural activity[8] for aesthetic judgements of beauty and 'goodness', even when people aren't explicitly thinking about either. This reflexive association may be the biologic trigger for the many social effects of beauty. Attractive people receive all kinds of advantages in life. They're regarded as more intelligent and more trustworthy, and they're given higher pay and lesser punishments, for instance.[9]

Brain-scan studies[10] reveal that the anterior insula, a part of the brain used for disgust and pain, is important in aesthetic judgements. This rather surprising result perhaps helps explain an evolutionary mechanism for our more general perception of beauty: that aesthetic processing is, at its core, the appraisal of the value of an object, whether something is 'good for me' or 'bad for me'. This appraisal is subjective and depends on the individual's current physiological state: a hungry person finds chocolate cake more attractive than a sated viewer. Our brain's aesthetic system may have evolved to enhance our value judgements about objects of biological importance, including food and mates, and then been co-opted by cultural evolution for things of social value, such as paintings and music. The scans revealed that our brain's responses to liking a piece of cake and a piece of music are in fact quite similar.

Our aesthetic appreciation probably also coevolved with our pattern-seeking impulse, which assists our brain's prediction system, perhaps as a cognitive signal to direct our attention: here is something special that needs figuring out or deciphering. Beauty is a motivational force, an emotional reaction alerting us to explore further – it is a particularly potent and intense form of curiosity. Art hijacks this instinct: If we're looking at a Van Gogh, that twinge of beauty in the brain's aesthetic centre is telling us that this painting isn't just swirls of colour, it is meaningful. In one brain-scan study,[11] scientists found that when listening to a familiar piece of Beethoven, the bits people found most sublimely beautiful were preceded by a prolonged increase of activity in the caudate, the same brain area involved in curiosity. This anticipation, the researchers said, signalling that a potentially pleasurable auditory sequence is coming, 'can trigger expectations of euphoric emotional states and create a sense of wanting and reward prediction'. The process triggers surges of the pleasure hormone dopamine. In a powerful way, beauty helps our brains work out which sensations are worth making sense of and which ones can be easily ignored.

So, we are biologically primed to respond to beauty and we culturally adopted this as a visual language: we made our objects of beauty into symbols with value and meaning. Humans see beauty everywhere and far beyond the charms of sexually attractive human bodies. The aesthetic pleasure we experience encourages us to spend valuable time absorbed in contemplation, to pay attention to objects of no practical or survival benefit, and to invest time, labour, and resources in our own creative expressions. No other creature does this. For such large animals, any unnecessary activity is particularly expensive. A spear is not going to do a better job at providing essential food if it is more pleasingly decorated. And yet the significant time, effort and material resources that all human societies devote to decoration speaks of an important survival role. It is through the symbolism and meaning of beauty that we draw unity, community, shared values and beliefs, compassion and other emotions that bind us as cooperative societies.

Our entire human world is built on symbolic renderings of thoughts and ideas, and it makes us unlike any other animal. We use visual symbolism to transmit our invented concepts between individuals and down through generations. Abstract ideas like currency systems, good and evil and government are expressed through beauty, using body decoration, art, music, architecture, gardening and other skills.

The foundations of our capacity for symbolism can be seen in our closest primate cousins. Youngsters from a group of well-studied wild chimpanzees, at Kibale National Park in Uganda, regularly play with stick 'babies' – found objects that they've imbued with meaning. Juvenile chimps in the group have been recorded cradling their sticks and carrying them into their day nests, something they don't do when the sticks are used for other play purposes. One young male made a separate nest for his doll, and a female was seen patting a log like she was 'slapping the back of an infant', which occurred while her mother was caring for a sick sibling.[12]

Some 2 million years ago, our hominin ancestors were also tending found objects. A cobble of red jasperite, with an obvious 'face' weathered into it by the elements, was found in South Africa alongside the remains of its archaic-human owners. The so-called 'pebble of many faces',[13] had been carried home from its most likely source, several kilometres away, and has been described as the earliest piece of found art. The stone face was valuable to someone long ago not because it was useful but because it had meaning. By the time of *Homo erectus*, people were deliberately beautifying their possessions – archaeologists have discovered decorated shells dating back 700,000 years in Java, Indonesia.[14]

Surely, though, the human urge to decorate, to communicate through symbolism, started with our own bodies. Every contemporary culture has a tradition of skin painting, whether it's reddening the lips or a more dramatic transformation, and ochres have been found at count-less prehistoric sites. Body decoration asserts an identity – it's a visual language, and can be used to show group allegiance.

The Ekoi, in the southeast of Nigeria, have developed an extremely complex form of group organization based on and expressed through decoration. Ekoi women traditionally have detailed symbolic tattoos on their face and body, including secret signs in the Nsibidi sacred language. These inscriptions, dealing with love affairs, warfare, or sacred elements, can be seen by everyone but only read by members of the Ekpe leopard society, which was the governing elite before col-onization. Explicit visual messages like these are just one of many ways people of every culture transform their bodies, in spite of the costs; they refashion themselves to outwit natural sexual selection. In this way, our culturally established appearances redraw those determined by our genes.

One of the most fascinating cultural experiments in human history is the invention of the personal ornament: the transmission of mean-ing to others. Since our deep ancestry, necklaces have been potent symbols, used to reflect cultural identity and social status, or to act as wearable tokens – small but powerful adornments believed to encourage vitality, fertility and wealth. Coloured shell beads worn by Neanderthals have been found in Spain dating back 115,000 years. The oldest necklace we've found for our species was discovered in Blombos Cave, on the southern tip of South Africa. At least 65 small, teardrop-shaped 'tick shells', with intentional perforations and traces of ochre decoration still clinging to them, reveal the common human-ity we share with their last owner, some 75,000 years ago. The Blombos beads would not look out of place hanging with the other necklaces on my cupboard. Whoever made the jewellery designed it with an eye for symmetry and beauty, carefully selecting the beads, and this mean-ingful intention was recognized by its wearer.

Beadworks are part of a decorative technology that conveys

information through a visual language shared by the wearer and their group, which may also be understood by a broader social network. Symbolic culture relies on collective belief. In my culture it is understood and accepted that my necklaces are worn for adornment – they would be interpreted very differently if worn in different cultures. Some cultures ascribe meaning to the colours of beads: Turkana nomadic pastoralists of northern Kenya give yellow beads to prospective marriage partners, while white is for widows. Anthropologists describe these shared beliefs as social norms, and they apply to everything from collective agreements on beauty to behaviour.

The archaeologists who analysed the many shell beads unearthed in the Blombos Cave realized that a striking change in fashion had occurred over time.[15] The wear patterns on beads found in the cave's lower, older layers revealed the shells had been hung free on a string with their flat, shiny sides against each other. But in the younger layers, the shells were knotted together two by two, with their shiny side up. This small, seemingly insignificant change in necklace style is the earliest evidence anywhere of a shift in social norms. It's the cultural-evolutionary equivalent of an anatomical distinction in the fossil record, or the refinement of a handaxe, only this time it's evidence of a new social adaptation. From such behavioural shifts we developed into complex, different societies, each characterized by its own identity.

It is not clear whether the earlier residents of Blombos changed their own fashion ideas, or if they were later replaced by another group of early humans who liked to wear their beads differently. But either way, these beads, like jewellery today, served a fully symbolic function, representing the social norms of their times.

Clothing also does this. Anthropologists argue that because we are an upright species, a clothing 'fig leaf' started as a social norm to hide sexual organs that would otherwise be continually on display: this action allowed large populations of unrelated people to live closely together without continual conflict. My own theory is that it originated with the practical use of slings to carry babies and loincloths for menstruating women. Like everything else we make or use, these would have become culturally significant, decorated, valuable and copied, perhaps by both sexes. Through symbolic dress codes, people signal status, gender and other culturally important messages, such as allegiance to their tribe and religion, which enhances the us-versus-them mentality that separates groups of people and augments social divisions within tribes. In this way, clothing plays a key role in helping distinct cultures develop, evolve and compete, each with their own technologies and expertise. And that's the culturally adapted evolutionary 'purpose' of decoration: to reflect social norms and bind members of a tribe together under a shared story.

The imitative nature of our cumulative culture means we copy our behaviours and preferences, so we are primed to seek out social norms and conform. Fashion norms can be as bizarre as they are impractical, but there have been ingenious subversions. In Japan, for instance, where non-royals were banned from wearing richly decorated silk kimonos, some women circumvented this by tattooing their patterns onto their skins. And decorative norms are entwined with other social norms, so, as the rights and status of women improved, their clothes became more practical. The invention of the bicycle hastened this emancipation, leading to the unthinkable in the form of trousers for women.

Through beauty we attempt to order our physical and social world, and make it conducive to our needs – we beautify ourselves and our made objects, but also our societies. Social norms extend beyond our visual decoration to govern our behaviour: we aim to look appealing but also to have beautiful manners. Social norms, which emerge in the group and are enforced by it, help explain how humans achieve our very high levels of cooperative activity. They evolved as an adaptation to reduce conflicts of interest between people by aligning our behaviour and values. As human societies grow, internal divisions and hierarchies emerge, and one strategy for avoiding conflict is through social norms that reinforce these divisions and remove other options. If a hunter's 14-year-old son is 'supposed' to join the hunters rather than the potters, because that's the 'tradition', then the whole group colludes to keep these roles clear, especially as many social norms are dressed in a narrative of supernatural ordinance that cannot be questioned. Rituals are often used to help bind unrelated individuals within sectors of society and strengthen social hierarchies. These include arduous and dangerous initiation rites, tests and ceremonies to tie people together through a shared experience.

Norms also help resolve conflict over shared resources. Most societies have rules regarding meat, for instance, with rituals and taboos concerning its preparation and over which bits of the animal may be eaten by whom. If a group of hunters brings back a kill, social norms dictate that select pieces will be reserved for, say, the person who made the arrowhead, the nursing mothers, and so on. This doesn't necessarily mean that the meat will be shared equally, but it means that everyone will get something and ensures that it is in the interests of the whole group to maintain the cohesion and uphold social norms.

In clan-based bands of hunter-gatherers, where there is a delay between order and delivery – the effort of producing food and the actual food itself – strict social norms help to avoid conflict. Take the Aché in Paraguay, who rely on forest-based beetle farming. First they must prepare a nursery by cutting down trees, but then they have a six-month waiting period until they can return to the area to harvest beetles from the felled trees. Very strict norms on property, therefore, govern trees that have been chopped down by a band. Inuit groups rely on similar rules regarding whale hunts (a dangerous but lucrative task): a speared whale rarely dies immediately, and it may take days or weeks to beach or float, by which time another group could claim it. Again, social norms dictate that whichever group's spear is in the whale when it arrives owns the whale.

Social norms are extremely powerful and dictate not just how we should behave in public but also in our private lives[16] – even when we are alone, as the surprising number of rules around masturbation attests. Norm constraints can stifle dissent and limit innovation, and may incur a personal cost – such as preventing Muslims from eating tasty bacon – but despite the inconvenience, we mostly adhere to the rules. Deviating from a social norm risks staining your reputation, and this also has an impact on your children, because in many societies, social punishments and debts are inherited by subsequent generations.

As with the fashions for clothing, many of our norms and rituals have no benefit: eating insects is considered revolting in my culture, but delicious in others. However, just as we associate visual attractiveness with goodness, so we also moralize with social attractiveness – norm-following behaviours are seen as inherently good, and by extension, so are their followers. In this way, our social norms maintain social cohesion by creating a shared basis for morality. They help us understand why people act as they do and, as a result, make it easier to predict people's behaviour.

Social norms govern our lives and yet they do not exist as objective properties of the world. Gravity exists whether you subscribe to it or not; murder is wrong in some cultural contexts, but medalworthy in others. This seems like a very obvious point to make, but so much of our human world is controlled by motivations, strategies and beliefs that we forget they are invented social norms – they become a part of what we think it is to be a human, and we accept them unquestioningly.

Take the roles of women and girls in society, which are more limited than men's. This is not because evolution has delivered big cognitive differences[17] – women are not intellectually inferior to men – but because social norms prevent women from taking prestigious roles. Patriarchal norms are so ubiquitous, you would be forgiven for thinking it was ever thus. In fact, sexual equality has been the norm for most of our evolutionary history, judging by anthropological and genetic data. Indeed, sexual equality and pair-bonding were two of the most important evolutionary changes to our social organization as we diverged from our primate ancestry.[18] Equality between the sexes offered a survival advantage because it fostered wider-ranging social networks (through both maternal and paternal routes) and closer cooperation between unrelated group members, facilitating trade in ideas and genes. Matriarchal norms may well have dominated in our ancestral communities. Today, hunter-gatherer societies remain remarkable for their gender equality, which is not to say women and men necessarily have the same roles, but there is not the gender-based power imbalance that is almost universal in other societies. In contemporary hunter-gatherer groups, men and women contribute a similar number of calories and both care for children. Men and women also tend to have equal influence on where their group lives and who they live with, increasing the opportunities for cooperation with unrelated individuals.

Although sex is biologically determined, gender is an invention of culture and often partisan. The majority of art is from the male gaze;[19] the world's major religions uphold and endorse patriarchal social norms. Most agrarian societies control and repress female sexual agency using norms that range from covering their bodies to murdering women who 'shame' them. Religions give spiritual authority to such practices, and often women pay the price of ensuring the group's favourable outcome. Thus, it is girl sacrifices that are found in the icy graves of Incan mountains, wives who must immolate themselves on their husband's funeral pyre, and daughters who were sacrificed in Athens. The cultural repression and control of females has had a profound impact on the social norms for women and girls to behave modestly and men to behave domineeringly. This is reflected in everything from Chinese foot binding to the enormous gender imbalance in life opportunities from health to wealth.

This cultural conditioning begins at birth, giving the individual the best chance of being accepted into society. Indeed, social norms even have an impact before birth: one study found that when pregnant women were informed of the sex of the fetus they were carrying, they described its movements differently.[20] Women who learned they were carrying a girl typically described the movements as 'quiet', 'very gentle, more rolling than kicking'; whereas those who knew they were carrying a boy described 'very vigorous movements', 'kicks and punches', 'a saga of earthquakes'. By contrast, women who did not know the sex of the babies they were carrying showed no such distinctions in their descriptions.

Many of the ideas we consider universally held are simply the social norms in our own culture. *Liberté, égalité, fraternité* may be values worth dying for in some cultures, but personal freedom is not considered important or desirable for many societies, which prioritize values like purity instead. Consider the idea of responsibility. In my culture, if you deliberately hurt a person or their property this is considered a much worse crime than if you did it by accident, but in other cultures,[21] children and adults are punished according to the outcome of their actions – intentionality is considered impossible to grasp and therefore largely irrelevant.

The danger of ascribing biological bases for all our actions, and not recognizing our social norms for what they are – culturally evolved motivations and behaviours that can be changed – is that individuals and groups are not given equal opportunities in life, and they suffer.

(Although whether you think people should be given equal opportunities is, of course, partly down to your cultural developing bath.) Social norms produce slavery, caste systems, 'honour' killings and many other damaging behaviours. It's worth noting, though, that many of those norms that were once believed set in biological stone or ordained by gods have been changed by societies – sometimes remarkably quickly.

Prejudicial social norms can become more egalitarian and vice versa. In the past few years, the social norm for judging someone by their skin colour or gender has shifted in the United States from being taboo to acceptable at the presidential level. There is no scientific basis for the belief that a person's skin colour or sex has any effect on their morality or intelligence beyond that imposed by society through social norms. And that is an important point, because social norms imposed on individuals and groups can change their behaviour and their biology.

Given that different cultures hold different truths to be self-evident, how do we come up with our social norms? It is a common misconception that they originate with a leader or depend on a centralized media source to coordinate a population. In fact, they seem to emerge spontaneously in societies. Take the fashion for baby names – in an online experiment,[22] anonymous players were randomly paired up and asked to agree on a name, before pairing with other players. At first it appeared that no winning name would ever emerge, as players suggested name after name, trying to match their latest partners' choices, with very little hope of success. Yet within just a handful of rounds, everyone had agreed on a single name. Consensus on the norm spontaneously emerged from nothing through random mixing, as changes in network connectivity enabled one name choice to take off, a concept that in physics is known as symmetry breaking. The results remained the same whether the game was played with 24, 48, or 96 players, suggesting that it would scale up indefinitely, and explaining how social conventions can spontaneously form even in very large groups like nations. The experiment also showed that the process of consensus building could be manipulated by adjusting how players interacted with one another, just as we saw for cooperators in the public-goods game. Simple changes to a social network make people more likely to spontaneously agree on a social norm: we like to conform.

But what about those of us who cling to individuality and don't want to go along with the herd – the teenager who feels alienated from mainstream culture, or the unconventional guy in his twenties who wants to make a statement? Both cultivate their look in opposition to social norms, perhaps by trying dramatic makeup or experimenting with hair or beard styles. But on revealing their radical statement, they discover that millions of their contemporaries have made the same choices and they all end up looking almost identical. This is dubbed 'the hipster effect', and mathematical signalling suggests that this kind of synchronicity spontaneously emerges as a property of large numbers of people. According to the models,[23] the majority of a group will conform to a norm, then there is a delay while the minority of nonconformers respond – their responses then undergo a phase transition to synchronize, producing the new hipster norm. In March 2019, after this research was reported by a technology magazine and illustrated by a stock image of a 'trendy' young man in a beanie, the editor received a furious response from a reader accusing the magazine of using a photo of him without permission, only to learn that the hipster in the photograph wasn't him at all – 'hipsters look so alike they can't even tell themselves apart.'[24]

Social norms bind us together in self-conforming cliques, helping us recognize each other as members even when we are unrelated genetically. Once an individual's fate is closely linked to her group's survival – and in competition with other groups – identifying members of your own group becomes vital. After all, our interests are best served by serving those who share our interests. Social norms about dress and adornment, behaviour, skills and practices all become ways to self-identify with your group and ensure you receive the help and protection that entails. This helps explain extreme beautification practices, such as cranial deformation, once used by tribes in Africa, Europe and South America, in which an infant's head was bound to a board for years to achieve the distinctive shape of its group. The more social norms a society has, and the more strictly they are enforced, the more predictable its members will be to each other – and the more suspicious they are likely to be of everybody else.

This is the genesis of tribalism. The more you share someone's social norms, the better you can predict their behaviour, so the easier it is to decide whether you can trust them to act in your interests – which lowers the costs of exchanges and interactions between people. From birth onward we learn consciously and subconsciously the social norms of our tribe, effortlessly 'belonging' to that culture by virtue of our birth and upbringing.

We can spot transgressors and fakers easily, from the politician who tries to cover his privileged background with a phony working-class 'street' accent, to the 'nouveau riche' social climber. Language, like cranial deformation, is an excellent tribal identifier because it is hard to fake – our ears are keenly attuned to pick up differences in accent, slips of grammar and nuances of phrasing that mark an outsider. A non-native who becomes proficient in a language may be able to communicate faultlessly, but it is unlikely they will be able to fool an in-group native that they belong. The challenges experienced by immigrants and other people who transgress our tribe's social-norm boundaries have been widely portrayed, including in George Bernard Shaw's *Pygmalion* and Jarvis Cocker's song 'Common People'. European Jews have been distrusted when they behave as an easily identifiable distinct minority, and trusted even less when they attempt to assimilate and adopt the majority cultural norms.

Another consequence of individual identity being so bound with the group's is that if someone switches tribes, she risks losing her own sense of identity and feeling alienated by both tribes, which has mental health implications (migrants have a higher incidence of schizophrenia,[25] for instance). But people will continue to try because being in-group is so valuable[26] for the protections and economic benefits that brings.

Tribalism – in-group/out-group recognition – is prehuman. Chimp groups are strongly tribal and very hostile to foreigners, with mortality rates from intergroup conflicts as high as 13 per cent. Unlike chimps, we live in larger mixed groups that aren't all kin, so we must confirm and assert our group identity and allegiance through cultural signifiers. Our out-group prejudice is learned in early childhood, and although we often dress up our animosity in terms of objections to cultural differences rather than the individuals themselves, in reality these are deeply held, cognitive patterns. By identifying people as members of an out-group, we clarify the parameters of the in-group and make our own position more secure within it. People feel a connection to other members of their group: their brains respond in empathy at their pain, for instance. But if they are told that another person is a member of an out-group, such as a rival sports fan, they stop emp-athizing.[27] When we look at someone we perceive as out-group, brain scans show that the neural firing pattern resembles the one we use for object recognition rather than people – we cognitively dehumanize them. Other research shows that the hormone oxytocin promotes altruism but only when people interact with someone they perceive as a member of their group – the same levels of oxytocin have no effect when they interact with an out-group person.

The whole premise of social cooperation within a tribe is that you can trust non-kin to act in your interests, so there is nothing more threatening to a group than the belief that someone is freeloading on your joint efforts or is simply not who you think he is. The more similar people look, and the more similar their cultural developing bath, the more important become the identifying tokens and social norms. Catholics and Protestants in Northern Ireland, like Hutus and Tutsis in Rwanda, look and sound the same as each other, and so they have had to accentuate any small differences. We look to our social norms to supply these tiny differences, whether they be rituals, taboos, religions, or foods. We craft a group identity through stories that cast us as the righteous side – as heroes or unfairly treated victims – in competitions with other groups.[28] These compelling narratives can be an extremely effective way of getting socially alike individuals to kill each other on the basis of belonging to opposed groups.[29]

It is when a group feels threatened that it binds itself most strongly in defence of its collective tribal interests – even five-year-old children act more cooperatively and generously with their group when it is under threat.[30] Groups of men who fight together are more likely to survive, and as generals know, each soldier is more likely to survive when the whole troop is prepared to die for each other. This is what's behind extreme 'proof of allegiance' type rituals in competitive disciplines. It also offers a cynical way to strengthen prosocial norms and institu-tions, and keep our societies cohesive: via competition and conflict with another group. This helps explain the rise of nationalism: its man-ifestation is indicative of a threat to the group, and this works as a feedback loop, convincing the group it is under threat from migrants or neighbouring countries. However, the threat in most of these nations is not external – they are experiencing an unusual era of peace; rather, the threat is from internal social divisions and inequalities.

Conflict between groups has always been a huge risk to life, with most hunter-gatherer groups experiencing repeated or continual conflict and a mortality rate calculated at around 15 per cent, similar to that of chimps. The mortality rate from conflict in the industrialized world is currently far lower, but in the past it has soared far higher, and the vast majority of conflicts have involved territory. Victorious groups expand at the expense of the losers, accumulating land, slaves, refugees and economic migrants seeking more successful economies. Intergroup competition probably drove many of our prosocial norms because only the most cooperative, cohesive groups made it. And the selection pressures within such groups increasingly promoted the most diplomatic people – those able to talk and charm their way past conflict and into favour.

Within these broad domesticating parameters, a flourishing of diverse groups practising different social norms emerged, and these differences changed the minds and bodies of the people who practised them. Cultural learning alters people's brains.[31] Practising any skill involves 'hardwiring' the neural networks involving muscle control and coordination, balance, judgement of speed and distance, and so on, until the skill feels automatic. Once a behaviour, action, or thought process is so practised it has become automatic, the brain's processing workload is significantly reduced, which frees up working memory, allowing the best of us to innovate the tiny details that push the boundaries of human excellence. This applies to everything from learning to walk to becoming a concert pianist or juggler.[32] People who played a lot of Pokémon as children share a specific brain region devoted to recognizing the game's characters.[33] Our cultural developing bath also affects the body – committed tennis players build up around 20 per cent greater bone density on the racket-holding side of their bodies; people who spend time at high altitudes produce more red blood cells and larger lungs to cope with the decrease in available oxygen. To be clear, these are not genetic changes, but biological changes during a person's life.

Those groups whose social norms and technologies improve health or economic outcome are more likely to survive and transmit their cultural practices to other generations. In many cases, this cultural evolution changes their biology. For example, a remote tribe of sea nomads, called the Moken, who live off the west coast of Thailand, have developed a unique dolphin-like ability to see underwater. Moken children spend much of their day in the sea diving for food, and their eyes have adapted to see twice as well as European children. Our underwater vision is usually blurry because the water refracts the light entering our eyes at the same rate as our water-filled corneas do, so we lose the ability to focus light. Moken kids, though, like seals and dolphins, have developed reflexive adaptations: their pupils constrict to the maximum known limit of human performance, thereby increasing the depth of field; and they change their lens shape. This is a biological adaptation to culture but it isn't a genetic change; it's a learned ability, even though it's unconscious, which means any child can adapt. Indeed, scientists tried training Swedish children to dive down and look at patterns on a card, and after 11 sessions, the Swedes had gained the same underwater acuity as the Moken children.[34]

However, for the Bajau people of Indonesia, another population of sea nomads, their culturally evolved lifestyle has produced a genetic adaptation. Geneticists investigating the Bajau's extraordinary diving ability have identified a handful of gene variants[35] in their DNA that allow them to hold more oxygen in their blood and important organs, control their carbon dioxide levels and also increase the size of their spleen, which acts as a reservoir for oxygenated blood – Bajau spleens are 50 per cent larger than the average. Some of these genes appear to have been inherited from our extinct Denisovan cousins during ancient sexual liaisons, and were selected in the population through cultural evolutionary pressures.

Our cultural developing bath profoundly changes the way we think, behave, and perceive the world. Studies comparing the neural processing[36] of Westerners and East Asians, for example, show that culture shapes how people look at faces[37] (Westerners triangulate their gaze over eyes and mouth, whereas East Asians centralize their focus) and how we perceive things in and out of context (Westerners tend to be good at isolating people and objects from their backgrounds, but bad at seeing them in context, whereas for most other cultures, the reverse is true). Asked to pick two related items from 'bus', 'train', and 'track', Westerners more often couple the objects of transport, picking bus and train, whereas Asians are more likely to pick train and track because one depends on the other. East Asian and Western people process information differently because of their society's different social norms, researchers[38] believe. Westerners, with an individualistic suite of social norms, tend to process objects and organize information into categories. In contrast, East Asians, with more collectivist norms, view themselves as part of a larger whole, and their processing prioritizes the relationships between objects and their backgrounds. In other words, the cultural developing bath creates biological brain-wiring differences. These differences reduce and become negligible the longer individuals spend in each other's cultures, and are wiped out by the next generation.[39]

Our social norms can have long-lasting genetic effects, however, because they impose restrictions on whom people have children with. For instance, in northern Thailand most societies have a social norm that involves the married couple moving in with the daughter's family or, more commonly, the son's family. This affects levels of genetic diversity – patrilocal populations experience a much larger influx of women than men, so such populations show little diversity in their Y chromosomes, which sons inherit from their fathers, geneticists found.[40] Where the norm is to live with the mother's family, however, people have varied Y chromosomes but show little variance in their mitochondrial DNA, which they inherit from their mothers.

Here's another striking example of how the cultural developing bath creates biological and behavioural differences. In a now classic experiment,[41] a group of American male students – half from northern states and half from southern – were told to fill in a questionnaire and deliver it to a table down the hallway. As each student walked down the narrow passageway, he passed a large man working at a filing cabinet, who needed to move out of the student's way. As he did so, he bumped the student and called him 'asshole' under his breath. By the time the student had delivered his questionnaire, he was either riled up with a blood-cortisol and testosterone spike to match, or he'd shrugged and laughed it off. The difference in individual response depended on which state the students came from. Most northerners were more amused than angered by the insult; 90 per cent of southerners experienced a flash of anger and showed a rise in stress hormones. If southerners subsequently met a stranger who had witnessed this 'humiliation', they acted in a domineering manner with a firmer handshake and reported feeling less manly in the stranger's eyes.

Southern US states have an honour culture with social norms that oblige and motivate men to defend their property, family, or reputations with violence. A relatively small slight, like being called a name, can spiral into full-blown aggression. In the next step of the experiment, the students who had just been insulted while delivering their questionnaires were confronted on their return down the narrow corridor with another man walking in the opposite direction, forcing them to step aside. Southerners who hadn't been insulted showed their good manners, stopping to step aside when nine feet from the stranger; northerners stopped six feet away. However, after being insulted, northerners waited another foot before stopping; southerners, though, didn't back down until they were about three feet away from bumping into the oncomer.

Stereotypes about groups usually have an element of truth. In the United States, southerners are, as a group, more friendly and polite than northerners, who are often more brusque and ruder. However, southerners have a stronger urge to punish than northerners – they are more likely to physically reprimand their children, and to approve of police using shoot-to-kill. These differences within a nation of people who ostensibly share the same language, environment and broad culture are not genetic but they are biological. Individuals' brains develop differences according to their specific cultural developing bath. The corridor experiment reflects regional crime statistics – FBI records show that southerners are far more likely to kill their friends and acquaintances in quarrels sparked by insults; the Deep South has twice the murder rate of the rest of the country. In other words, an individual's cultural developing bath influences their survival.

Honour cultures arise in places where people's resources are vulnerable and government is weak – places where dominance rather than prestige is the selection pressure on social norms. Around the world, these are typically remote herding societies prone to livestock rustlers, where there are few opportunities for cooperative collaboration and a reputation for violence – for taking no shit – is necessary protection. Most violent acts are provoked by attacks on a person's honour and are preceded by feelings of shame or humiliation. Agricultural societies, by contrast, where large populations of settled people live closely and must cooperate over land sharing and common infrastructure such as irrigation channels, tend to result in social norms that value prestige over dominance. Crops are not a profitable steal in the same way as cattle, and farmers can rely more on stronger collective-action institutions to punish wrongdoers than self-defence. Instead of aggressive acts to deter neighbours from attacking you (in which you risk injury), generosity and cooperation toward your neighbours enlists them to help protect you from dangerous situations.

The southern United States was populated by Scottish and Irish immigrants, moor and mountain herders who brought their culture of autonomy and honour with them. In many places they settled, they assimilated into local farming or urban culture, but in the remote, rural Deep South, their 'every man for himself' honour culture persisted. The northern states, by contrast, were settled by crop-growing German and Dutch immigrants with strong community institutions. Social norms are resistant to change because people don't invent their attitudes, they learn them from their parents.

Ultimately, most social change is driven by economics. Europe's honour culture of duelling among the aristocracy died with the rise of the middle classes, which made the practice look ridiculous by demonstrating sensible ways of resolving disputes, and stronger social institutions meant duellers were liable to be charged with murder rather than celebrated for defending their honour.[42] More recently, a shift has occurred in the very strict honour culture of Yazidi herders in the aftermath of atrocities by Islamicist militia against the tribe's women. Thousands of Yazidi survivors of abduction and rape feared returning to their villages, because of strong social norms ostracizing women whose sexual reputations have been damaged. However, the economic and social necessity of allowing the survivors to return to decimated villages prompted deliberate change. Women were offered a ritualistic 'cleansing' opportunity and accepted back into the community (often gaining new freedoms), demonstrating a way for everyone to save face and the community to heal after an atrocity. One of those women, Nadia

Murad, was globally recognized for her courage and earned a Nobel Peace Prize in 2018.

Honour cultures are slowly dying out – intimidation is a barrier to social cohesion and so such societies often disintegrate and fall prey to more prosocial groups. The trend is towards more prestige-based cultures, like those in the northern US states. Meanwhile, as populations become more diverse, and people are exposed to a range of different norms, such as in cities, tolerance of transgressors increases, so a broader diversity of individual expression emerges. Being exposed to different social norms, especially from a young age, makes people more open-minded – studies show that when children attend more ethnically diverse schools, it leads to greater social cohesion between ethnicities.

Social norms and their decorative expression arise from the collective belief systems of a tribe, but they also in turn influence their society's cultural beliefs and identity. For instance, homosexuality is widely practised across North Africa and the Middle East, and it is tolerated by Islam.[43] However, while private homosexual activity has been a social norm for many centuries in the Arab world, the relatively recent Western norm for expressing homosexuality is strongly rejected: men waving rainbow flags have been thrown into jail in Egypt, and a school in Riyadh was fined $26,000 simply for having a parapet painted in rainbow colours, 'the emblem of the homosexuals'.[44] It's not that same-sex coupling is any different in the two cultures, but the social expression of it is entirely different. The vast majority of Arab-world men partaking in same-sex sexual acts would not describe themselves as homosexual; in the West, many more of them would. The prevailing social norms dramatically influence decoration, because it is a signifier of identity.

Through beauty, we generated a visual language that enabled larger groups of people to cooperate together as a tribe, united in a shared identity, social norms and collective belief system. Operating on this scale brought energetic, economic and survival advantages and enabled us to compete with other tribes for resources. Yet the great paradox of human culture is that while we are primed for tribalism, we rely on networks of cooperation between our tribes to exchange ideas, resources and genes, as we shall explore next.

10

Trinkets and Treasures

In January 1492, a solitary man rode a mule out of Córdoba, Spain. Around him, the decaying grandeur of Europe's once finest city mirrored his own dejected state: past his prime and with little hope. He had spent the best part of a decade pursuing a wild dream but with his latest campaign for funds rejected, the 40-year-old mariner was forced to concede defeat.

Cristoforo Colombo had been born the son of a weaver in Genoa, a cosmopolitan port city that, hemmed in by mountains, looked outward across the sea. It was easier to get to Lisbon than to the closer cities of Milan or Geneva, and Colombo spent most of his adult life exchanging goods by ship along the Atlantic coast between Portugal and West Africa. By far the most important commodity of his time was spices. High prices, a limited supply, and their mysterious Eastern origins had fuelled a lucrative European market, but with the mighty Ottoman Empire controlling all roads to the East, the spice trade had became increasingly dangerous and costly.

A sea route was sought. In 1488, a Portuguese mariner had made the first successful sailing around the southernmost tip of Africa to the Indian Ocean, but it was a perilous course. Colombo had an alternative proposition, to sail westward from Europe and reach Asia directly, thereby avoiding the stormy cape. During his childhood, the printing press had been invented, and Colombo read widely, forming the opinion from his studies that the circumference of the Earth was some 20 per cent smaller than generally accepted. However, his many petitions for expedition funds were turned down by the kings of Portugal, Genoa, Venice, England and, finally, the Spanish king and queen, Ferdinand and Isabella, all of whom rejected his 'small Earth' theory. Then, seemingly on a whim, Ferdinand changed his mind, and Isabella dispatched a royal guard to chase after the departing mule-rider. Colombo was given an annual allowance and promised various rewards in the unlikely event of his success.

That October, Christopher Columbus landed in the New World, ending a 10,000-year isolation of the American people (one-third of the world's population at the time) and starting a process of globalization – of interconnected interdependence – that would transform our world. The Columbian Exchange brought silver, gold, minerals, foods, tobacco, syphilis and turkeys to Europe, and thence to Asia and Africa. To America, it brought diseases, slavery, extinctions, Christianity, livestock, guns and people. The impact was rapid and extreme. Advanced American civilizations were devastated within decades, as 90 per cent of the indigenous population was extinguished by measles, smallpox and flu. Over 3 million people perished on the island of Hispaniola[1] alone, under Columbus's own cruel reign.

Meanwhile, for Europeans, the trade in resources and enslaved American and African labour reduced the energy costs of innovation and funded a cultural explosion in creative ideas, technologies, architecture, arts and trade. Bolivia's Cerro Rico mountain alone yielded some 70,000 tons of silver,[2] enough to bankroll the Spanish Empire for more than two centuries. The new wealth pouring from the Americas into the European elite restructured and strengthened institutional hierarchies, enabled Christianity to push Islam out of Europe, and ramped up exploration, trade, colonization and private enterprise across the known world. The Dutch and English especially profited from the control of the spice trade in the East Indies,[3] particularly the Spice Islands, which were the only source of nutmeg and cloves. Wars were fought, lands colonized and fortunes made.

The consequences were truly global. The flourishing of industry and creative expansion in the advancing Western economies occurred at the expense of the stifled economies in the global south, whose populations were impoverished, resources stripped and their cultures consequently – or deliberately – crushed. Cultural knowledge acquired over generations was lost as demographics changed; tribes were split and relocated or were no longer allowed to practise social rituals. In some cases, new migration swamped indigenous populations and the new culture and language took over. In others, populations died as a result of disease, conflict, or famine. Today, a generation after Western colonialism largely ended, and decades into our modern globalized economy, the cultural and economic effects are still stark and entrenched.

Yet when Columbus died in Spain in 1506, rich from hijacked gold, he never knew what he had discovered – all he thought he'd found was outlying bits of Asia.

The root cause of this planetary-scale cultural, environmental and genetic exchange was the desire for spices. It was this that helped fuel European colonial empires to create political, military and commercial networks under a single power. And yet spices had an entirely arbitrary and invented value – the word 'spice' derives from the Latin 'spec', the noun referring to 'appearance'. Spices were desired for their beauty; colourful, exotic, aromatic and flavoursome, as a nutritional resource they were essentially useless, and any alleged preservative benefit was outweighed by the fact that fresh meat was cheaper and easier to obtain than spices. In other words, pepper, cloves, cinnamon and nutmeg were desirable because of the cultural value we ascribed them. Once this value was accepted by society, spices became objects of conspicuous consumption, a mark of elite status and the first globally traded plants. The spice trade was a phenomenally important global activity because the human quest for beauty is so powerful.

Aside from its role as a signifier of tribal belonging, beauty plays another key role in human culture: we use it to ascribe meaning – and, therefore, social value – to *things* irrespective of their survival benefit. We value beauty: rare flavours, like spices; difficult-to-make colours, like purple; shiny materials like silk, precious stones and metals. We delight in the very uselessness of ornament. Long before the Columbian Exchange, our ancestors were harnessing this innate desire for beauty to lower the costs of trade, building networks that boosted cultural complexity and improved survival. Trade was the cultural lever that enabled our species to compete through cooperation, by exchanging resources, genes and technologies to expand across the globe – beauty was its facilitating mechanism.

The earliest human societies, like many small societies today, traded with each other through barter relationships. Although each group draws strength from its own patriotism and prejudice against outsiders, in reality, they operate in a web of dependence, much like the individuals within them. Tribes cooperate over resources, in collaboration against other tribes and in the trade of skills and materials. Indeed, trade is so important that some anthropologists believe it may have driven the emergence of language, without which even straightforward exchanges are difficult. People trade voluntarily because both sides believe they gain more from the exchange than investing in generating everything they need themselves, and they usually do – just as specialization makes sense within a tribe, it also makes economic sense to specialize as a tribe, as the nineteenth-century economist David Ricardo showed.

In Ricardo's scenario, there are two countries – one is good at producing food, and very good at producing clothing; the other is bad at producing food and very bad at producing clothing. You may think that the first country, being better at both tasks, should make both food and clothing, and ignore the second country. In fact, Ricardo showed mathematically that it is most efficient for each country to specialize in what it is best at and trade with the other – comparative advantage is more important than absolute advantage.[4] We trade with each other because it improves our survival: specialization is the most energy-efficient strategy,[5] which is why it is ubiquitous across biological systems, from ants to brain cells. Specialist skills, such as spearhead manufacture or whale hunting, developed based on barter between groups, enabling cultural practices and technologies to increase in diversity and complexity.

If one group has no whale meat but makes spearheads, they can trade with a hunting group that needed this tool. But what if the hunters have no meat to trade with and need the spearhead in order to get it? Just as we saw with the bison cloak, this sort of delayed reciprocity requires trust, since it involves one party handing over a commodity in the hope that they will eventually receive their compensatory meat. While skills improve faster through specialization, they also become ever more dependent on each other, without the social norms and reputational controls that exist within groups. And it's easy to see how exchanges can become more complicated. What if there are no whales around but there are sweet potatoes? Sweet potato gatherers have no use for a spearhead, but spearhead makers still need to eat.

Barter relies on a coincidence of supply, skills, preferences and time. Most of these issues can be solved in very small societies, but barter doesn't scale well. As societies grow bigger and networks become more complex, involving multiple exchanges relying on trust between strangers, it becomes difficult to keep track of goods and services and disproportionately costly. Relying on reputation and social norms to safeguard delayed delivery, whether 'owed' by nature or in trade with others, becomes risky. Reputational beliefs can suffer from errors about which person did what and valuation disagreement. Over time, the mental load of calculating transaction costs and risks becomes a problem, hindering trade and potentially leading to conflict.

Beauty – in the form of desirable items – solved this problem. Humans are acquisitive, we have a collecting instinct[6] – like bowerbirds and magpies, from childhood we collect things that we desire simply for their beauty, and our cultural evolution hijacked this urge. By age three, children have a strong sense of ownership[7] and will object to their possessions being replaced, even with identical versions.[8] As societies got big enough, private property norms helped transform us from people who decorated ourselves to people who owned decorated things. The transfer and exchange of collectibles replaced reputation as the enforcer of reciprocation between tribes. Trade boomed.

Take the ancient shell necklaces in Blombos Cave – part of what makes them so special is that they are collectible. Finding the shells and making the necklaces must have had an important selection benefit since it was costly – manufacture of the necklaces took a great deal of skill and time when humans lived on the brink of survival. A compelling theory is that aside from their social purpose in strengthening tribal identity, these beautiful trinkets were collectibles that could be exchanged: the first form of money.

The same perforated shell beads at Blombos have been found at sites from Algeria in north Africa to the southern Cape and across to Israel, dating back 120,000 years,[9] suggesting this was a cultural practice shared across tribes over many thousands of years. Several of the locations where the marine shells have been found are so far inland that they must have been brought there, revealing active, continent-wide trade networks between coastal and inland peoples. The shell beads probably helped create and maintain these networks. These ancient organized networks would have also enhanced genetic and cultural exchange, and thus accelerated our cultural evolution. For, just as we evolved to rely on our tribe for survival, so the survival of our tribe is reliant on other tribes – trade networks were vital for our African ancestors, just as they were more recently for ice age Australians. (To recap, this is a key difference between cultural evolution and genetic evolution: although there is some group selection in biological systems, it's debatable to what extent;[10] in cultural evolution, however, group selection is a big driver through reputation and social norms.)

The value of collectibles helped drive the development of skills and technology to make them, as well as the trade in and exploration for resources. Beauty became an important tradeable resource, satisfying a cultural 'hunger', as well as playing its part in lowering the transaction cost for trade in resources to satisfy our biological hunger, such as food or territory. Valuables could be used as security during exchanges with delayed compensation, gifts to compensate in-laws for loss of their child at marriage, or trophies to placate a hostile tribe. Some collectibles also give authority to a societal position: a crown marks a chief, for example. These are usually passed on to the next titleholder, but they carry their own power and can be used to give authority to a usurper in conflict. Upon death, collectibles can be distributed among heirs as inherited wealth – we are the only animal to have 'wealth' – but also as titles with privilege and responsibility. This means that we acquire not only a genetic inheritance from our parents but a sociocultural one too, all of which affect the survival of our genes (and cultural knowledge) over generations.

When someone has deliberately beautified something, it becomes heavy with meaning, and we recognize and value meaning even if we can't interpret it. There is a curious little gilt Celtic cross held in the British Museum's brooch collection, which was found in the Ballycotton bog in Ireland. It dates to the eighth or ninth century, and what makes it special is the small glass jewel embedded in the centre, on which is inscribed 'in the name of Allah', in Arabic script. The nearby port in the west of Ireland was, at that time, a significant trading centre, and it's likely that someone from the Muslim world dropped the jewel. It is unlikely that the person who found it, 12 centuries ago, would have been able to read at all, let alone in Arabic, but he nevertheless recognized it as a deliberately made collectible with meaning, and, therefore, of value. And so it was mounted into another symbolic piece.

For most of human history, we have lived as nomadic hunter-gatherers or herders, owning the bare minimum of stuff that can be packed up and carried. The few personal items such people own, usually jewellery and perhaps a decorative textile such as a rug or item of clothing, tend to be valuable and collectible. Today, Turkana herders treasure their beaded necklaces, and Mongolian herders may keep textiles and an intricately decorated 'ger' door – trade in these help with economic uncertainty in an unpredictable life. Objects valued as collectibles act as insurance policies because they are economically significant, and this in turn spurs the development of decorative material culture.

A small museum in the Bavarian town of Ulm, on the banks of the Danube, holds in its collection an exquisite figurine, called the Lion-Man. Carved 40,000 years ago from a piece of mammoth ivory, this mysterious beast has the head of a cave lion – its carver's most fearsome predator – and the stance of a human, and is the oldest known physical representation of a supernatural being. LionMan stands just 30 centimetres tall, and yet, with its masterfully carved body, attentive face and direct gaze, it is an incredibly powerful object. Experiments show[11] that this piece of decorative symbolism would have taken a skilled person more than 400 hours to make, and the wear on his body reveals repeated handling over time. LionMan is a beautiful ornament that must have held spiritual significance for the society that made him. Perhaps he was a deity that was able to pass between the human and animal worlds.

This ice age creation, like the shell beads, is a piece of raw material that has been made valuable through beautification – meaningful decorative expression – rather than because it fulfils a biological need. The society that made this figurine valued creative skills and were able to devote the time and human resources to learning and practising them. It was found carefully stashed with other collectibles – perforated arctic fox teeth and reindeer antlers – in an inner chamber of a cave complex. Inside the LionMan's mouth are the microscopic remains of something organic – the archaeologists suspect it is blood. It is tantalizing to think of this manufactured symbol playing a significant part in the collective narrative of this sophisticated ancient society, drawing them together as a tribe strong enough to survive ice age conditions, cave lions and their human competitors. The decorative objects and evocative paintings these first Europeans left tell of a creative, resourceful people, who didn't simply survive some of the harshest conditions[12] in human history, but thrived, outcompeting their Neanderthal cousins by using strong trade networks to exchange culturally learned ideas, technologies, resources and genes.

When modern humans displaced Neanderthals, population densities increased at least tenfold.[13] The main way they increased the land's carrying capacity may well have been through wealth transfers made more effective, or even possible, by collectibles. Neanderthals also made a range of decorative items but it is not clear that they traded widely. Our ancestors, however, collected and traded raw materials over great distances, and from them made musical instruments,[14] figurines, jewellery and other decorative items of added value that they traded. Trade allowed us to build greater social networks, increase our group size and cultural institutions, and our resilience to the harsh environment. It enabled our ancestors to occupy land across continents, whereas Neanderthals never ventured past Eurasia.

Hunter-gatherer tribes often separate into bands during the hunting season, and then come together en masse at great festivals for a week or so, a few times a year. During these gatherings the various craft makers and specialist hunters from different tribes and cultures all interact. Meat, stories and other resources are exchanged, ideas, techniques and tools are pored over, dances, music and decorative items are examined, and trading relationships are developed. In preparation for these festivals, modern-day hunter-gatherer societies, like the !Kung peoples of the western Kalahari, devote substantial time to preparing and manufacturing tradeable collectibles, such as ostrich-shell jewellery. This is a valuable investment in the group's time and energy – one of the main things the !Kung buy with their collectibles is rights to enter another group's territory to hunt or gather food there. Collectibles enable them to benefit from this survival arrangement – an insurance policy for hard times.

Our ancestral African tribes used collectibles to expand and travel. Trade helped ancient groups to migrate because it spreads environmental risks – if the water holes dried in one tribe's territory, causing a dearth of game, then it would be possible to acquire food in an exchange from another tribe farther off. Migration is an adaptation that allows humans to survive changing environmental and social conditions, but entering another tribe's territory is dangerous, and in doing so our ancestors' behaviour marked a big break from primate behaviour. Chimps, for instance, regard all members of another group with hostility and attack any intruders. When chimpanzees expand their territory, they do so by attacking and killing their neighbours. Humans also take territory by force, but we frequently use other techniques, such as diplomacy, that allow tribes' safe passage or enable them to share territory or to buy it through trade. And when a tribe is conquered by force, the losers are not always slaughtered; they might be made to pay tribute, be enslaved, or forced to switch allegiance to the victorious tribe by following its norms, which thereby benefits the conquerors in labour and resources.

One reason human intergroup interactions are often cooperative rather than hostile is because of our relatedness, which makes trade and migration through neighbouring territories easier. Our extended families, including in-laws, often straddle group boundaries. But the main reason humans don't regularly annihilate their neighbours is because of the far greater benefits of doing business with them; as a consequence, we developed social strategies for intergroup interaction. Through friendly language and other signals of good intent, including offering collectibles for right of passage, we can approach a strange group and it is likely we will be unharmed. Most societies have social norms that welcome strangers, and the reputation of the group and the prestige of its leaders hangs on ensuring visitors are shown good hospitality and leave with an impression of the tribe's generous behaviour, manners and wealth. This smooths the path to trade and means that ideas can spread. The benefits can be seen in trading blocs, such as the European Union, which have effectively made the costs of competitive conflict prohibitively unattractive compared to peaceful cooperation.

While all other primates are limited to tropical forests, human trade networks bridged tribal barriers and enabled ideas and people to flow geographically, boosting the diversity and complexity of culture, and also changing our environment and genes. By mapping the appearance and frequency of genetic markers in modern peoples, we can map when and where ancient humans moved around the world after they left Africa, probably crossing the Bab-el-Mandeb Strait from present-day Djibouti to Yemen. Some of these early beachcombers expanded rapidly along the coast to India, and reached Southeast Asia and Australia 65,000 years ago.[15] Meanwhile, another group appears to have set out on an inland trek from the Arabian Peninsula, through the Middle East and across to south-central Asia. From there, tribes colonized the northern latitudes, and there is evidence of them in China 80,000 years ago,[16] Europe by 40,000 years ago. Finally, during the Last Glacial Maximum, around 20,000 years ago, with sea levels 90 metres lower than today, a small group of Asian hunters headed into the frozen East Asian Arctic, crossing a glacial land bridge[17] to the Americas. It took another 5,000 years to reach ice-free land to the south, and less than 1,000 more to make it all the way to the tip of South America. In so doing, our species of tropical ape occupied all the continents except Antarctica.

The hostile environment of the Pleistocene era, where humans have spent most of our evolved history, kept populations low and limited trading opportunities between separate groups, and this is reflected in genetic differences that subsequently arose between separated descendants of the same relatively small group of African explorers. Thus, Tibetans, whose ancestors first colonized the high plateau around 25,000 years ago, have overcome altitude limitations of most placental mammals with a gene (inherited from ancient Denisovan matings) that helps pregnant women manage low blood-oxygen levels. Tibetan women with the gene have more than twice the number of surviving children as those without it, indicating a very strong selective force. In the Andes, another high-altitude region that was first populated some 11,000 years ago, people have undergone different genetic adaptations,[18] raising the concentration of haemoglobin in the blood and improving how it concentrates oxygen. Skin colour,[19] controlled by several different genes, is an obvious marker of ancestral migration, generally lightening (losing melanin) with latitude, as the sun's strength weakens. Melanin protects against ultraviolet rays, but limits the amount of essential vitamin D the skin can make in its reaction with sunlight. The familiar pale-skinned European is incredibly recent, though.[20] Europeans had dark skin and hair[21] even 7,000 years ago, according to genetic analysis of Spanish hunter-gatherers.[22]

Europeans owe their pale skin, languages and much else to a remarkable people who established the world's first transcontinental trading network and, in so doing, overhauled the genes and cultures of tribes from Ireland to China. Some 5,500 years ago, this highly successful tribe of nomadic pastoralists, the Yamnaya, journeyed north from territories along the banks of the Black and Caspian Seas in the Eurasian steppes. They had magnificent merchandise – and also the logistics for it. The Yamnaya's own transformation had begun with domesticating – rather than simply hunting – wild horses, which gave them a beast of burden and a war vehicle. Then they invented the wheel,[23] allowing them to travel farther and faster while transporting goods. When drought hit their grasslands they sought greener pastures and new trading opportunities – some hitched their wagons for central and northern Europe, others ventured east into Asia.

The Yamnaya would have been an extraordinary sight, like nothing European crop farmers had seen before: fair-skinned, dark-eyed warriors decorated with bronze jewellery, who rode horses that pulled wheeled wagons. These were people who spoke an Indo-European language and had advanced metalworking skills, who made collectible decorative jewels and intricately patterned bell-shaped earthenware known as Beaker pottery. These fashionable pieces were widely traded, and have been dug up from Scandinavia to Morocco. According to a recent analysis,[24] the Yamnaya also smoked cannabis and set up the first Eurasian trade in marijuana.[25]

The Yamnaya were successful herders and breeders, transforming the wild cattle, goats and sheep into new species of domesticated docility to provide food, hides, blood and milk products. Many herders harvest blood from their animals – a useful way of acquiring calories and protein from living animals – but the Yamnaya may have been the first to also milk their herds. Many of the Beaker pots contain remnants of milk, and it's likely they were making yogurts, curds, and cheeses, just as nomadic herders on the steppes continue to do today. This culture changed their genes.

Although mammals survive on milk during infancy, the gene that allows them to digest the lactose[26] in milk ceases production after weaning, and by adulthood they can no longer drink it.[27] Yogurts and hard cheeses contain very little lactose, so they don't pose much problem, but these first herders also experimented with unprocessed milk – and their genes responded. Around 9,000 years ago, a genetic mutation emerged in the Yamnaya that allowed older children and adults to digest milk. Those people who inherited the lactase-persistence gene would benefit from sugar, protein, fat and other nutrients in milk; whereas those without it would have become weakened through trying. Any adaptation that improves nutrition spreads quickly through a population, because healthy people are more fertile, more likely to survive childbirth and more likely to pass on their genes. We took a wild auroch and guided its evolution to produce a domesticated cow. We drank its milk and our genes adapted. This is cultural-environmental-genetic evolution.

Within a couple of centuries, the Yamnaya had revolutionized European society, culture and genes, driving Stone Age farmers rapidly into the Bronze Age. Lactase-persistence would have been a great advantage to stunted and malnourished crop farmers: today, around 98 per cent of northwestern European adults can drink milk.[28] Fair skin, too, would have benefited crop farmers with little access to animal liver and other food sources of vitamin D,[29] surely contributing to its spread. In societies with small populations, even slight advantages can lead to a gene proliferating. Similarly, the social norms, institutions and technologies of this advanced tribe were copied and adopted by other groups. This is how belief systems, jewellery, arts, technologies and institutions of successful tribes spread across large areas. Each tribe then stamps its own identity on these: the result of this cultural, genetic exchange was a predominantly fair-skinned, lactose-tolerant people, who spoke a new proto-Germanic language (Indo-European languages with adopted agrarian terms); practised crop, livestock and dairy farming; and invented a new pottery style, known as Corded Ware, made by Stone Age women potters evoking the decorative wooden caskets of the Yamnaya, to make big beer-drinking vessels.[30]

The Yamnaya were so transformative because they were networked – a web of mobile societies that formed an intercontinental communication system, using faster transport with horses and wagons carrying food and water – and because of their trading prowess. They were surely helped by timing, arriving in Europe shortly after plagues had swept through the continent and devastated the population.[31] Bands of male[32] Yamnaya warriors, wearing dog- and wolf-tooth necklaces,[33] stormed through Europe on horseback, colonizing villages. The indigenous men were overwhelmed, slaughtered, or evicted – DNA evidence shows that the last refuge of these first farmers was Sardinia[34] – and the women were raped or chose to partner with these tall, healthy, exotic foreigners. Around 90 per cent of the original gene pool was wiped out by the Yamnaya,[35] including all of the men in what is now Spain and Portugal.

These Bronze Age pastoralists were the pioneers of globalization, exchanging food, knowledge, metalworking technology and cultural skills across vast stretches of Europe and Asia. Some of what they traded was useful, such as metal tools, but much was purely decorative – beautiful objects that circulated widely and thereby oiled a broader, more efficient economy. The trade routes created by Yamnaya and their neighbours[36] continued to be important exchange networks for valuable collectibles, including amber, silk and spices, eventually becoming part of the Silk Road several millennia later.

The Yamnaya were genetic and cultural revolutionaries at a time when global population was just 5 million. By the height of the Silk Road, we numbered as many as 360 million people, and greater population meant greater cultural and genetic diversity. Trade routes had become complex networks that didn't simply export and import cultures, but whose interactions propagated new ideas, technologies and beliefs, accelerating cultural evolution.

Arguably, the Silk Road's origins began well before Eurasian steppe people had figured out how to capture and domesticate a wild horse.

Some 7,500 years ago in China, artisans began taming a far tinier creature: *Bombyx mori,* the silkworm. Over several centuries of breeding, they produced a domesticated version, which was larger, bred faster, produced more eggs and managed a tenfold higher silk production. The artificial silk moth could no longer fly, so it was entirely dependent on humans for reproduction and to supply its diet of mulberry leaves. Its larval stage, the silkworm, was edible, but the real prize was beauty: the incomparable silk that the worm produced to cocoon itself during metamorphosis. This could be spun into a lustrous, strong, yet fine material of high value.[37] It was at one time so valuable that a length of silk cloth was used as a monetary standard for exchanges and peace offerings with tribes, and even to pay soldiers and other workers. Human-caused ecological change – our directed evolution of a wild species – had produced something of no direct biological benefit but to which we ascribed a cultural value.

This decorative string, made by an unremarkable-looking worm, was China's most valuable commodity, and it changed the world. From Egypt to Rome, people lusted after the fine cloth and sent spies to try to decipher the secret of its production. By the second century, the Yamnaya's ancient trade routes had been broadened into a 4,000-mile network between the Pacific and the Mediterranean, a vitally important economic and intellectual link between cultures that persisted for centuries, connecting previously isolated populations. Everything from Buddhism and Islam to spices, precious stones, metals and ceramics travelled along the Silk Road, including, less desirably, the bubonic plague. Drought pushed flea-infested marmots and gerbils from central Asia carrying the deadly *Yersinia pestis* bacterium, which infected merchant caravans along the route. By 1345, plague had reached Black Sea ports, and from there spread to Constantinople, the Middle East, Egypt and the Mediterranean. The dying was unimaginable: in Europe, nearly two-thirds of the population perished – up to half of Londoners were wiped out; in parts of East Anglia, seven out of ten died. With formerly prosperous cities from Norwich to Florence deserted, it must have felt like the end of the world.

It was certainly the end of a world order. Disruption to a network – such as from epidemics or war – helps shake people out of 'safe' ways of doing things and allows new connections to be made; different people, ideas and technologies to become prioritized; and new network shapes to form. The societal restructuring that followed the Black Death facilitated the rise of the Ottoman Empire, making the Silk Road an expensive, dangerous passage for European traders, which, as we have seen, prompted the discovery of the Americas.

Eventually, the secret of silk got out. Although whether it was via a Chinese princess who, on marriage to the Jade King of Khotan, smuggled the silkworms and mulberry seeds in her turbans, or whether it was two Byzantine monks who smuggled silkworm eggs out of China inside their bamboo canes, we will never know. The Chinese lost their monopoly on silk production, although they continued to be major exporters, but silk had catalysed a cultural and genetic exchange of tribes who had been separated by mountains and rivers and thousands of miles.

The great mixture of genes, people, cultures and technologies that exists today owes much of its variety and complexity to beauty – to the human propensity to invent a value for objects that are not biological necessities but which we nevertheless desire. The benefits of trade forced us to cooperate with other tribes with different social norms, genes and technologies. It thus expanded our networks – and our collective brain – and encouraged us to explore our environment in search of valuable raw materials. Trade boosts cultural evolution because technologies and behaviours that have been nurtured and selected in a tribe are brought under new types of selection pressure by other populations – a meta-selection process that increases cultural complexity and diversity. In some cases, new ideas and technologies were exchanged along with resources and collectibles; in other cases, the people themselves were part of the exchange, through migration or other means, and their culture has replaced an earlier one. We can see the differences using population genetics, but either way, for cultural complexity to increase, history shows that populations and social networks need to increase.

Social networks create synergy, allowing configured groups to achieve things that a disconnected collection of people could not. Columbus only managed to bring about such a dramatic cultural exchange because he was part of an organized, international trading network. Throughout human history, where networks were strong and broad and the climate was conducive, people made fancier technologies. When the reverse happened, cultural complexity was lost – sometimes for thousands of years.[38] This helps explain some of the cultural gaps in archaeological finds that are not the result of poor artifact survival.[39]

Where societies become isolated, cultural (and genetic) complexity can drop to the extent that they go extinct, or struggle to survive. This is what happened to Tasmanian aboriginal people. By the time Europeans arrived on the island, it had been separated from mainland Australia for at least 10,000 years. The people living there were struggling in small isolated populations and their technology had simplified to the extent that they had just 24 different tools, including badly made, leaking boats, and they no longer fished – they had also, it is claimed, lost the ability to make fire, due to their cultural and economic isolation. Compare this to Pama-Nyungan-speaking Aboriginals, who lived directly across the Bass Strait from Tasmania, and who at the same time had hundreds of complex and multi-part tools, boats, specialist clothing and a variety of nets and spears for fishing and trapping birds and other animals. The Tasmanian tool kit was cruder by far than those used by Europeans 40,000 years ago, and it was certainly cruder than the ancestral technology they had before leaving mainland Australia. The isolation of the Tasmanians had effectively shrunk their collective brain – their cultural evolution had simplified.

A similar thing seems to have happened with Canadian Eskimos. These resourceful, skilled reindeer hunters with impressive, specialized tool kits were part of an Arctic-adapted tribe, a small group of whom navigated their way across the hostile ice and sea from Siberia to North America, around 6,000 years ago.[40] For 4,000 years, they survived Arctic climate changes, including cold spells that depopulated most of the region, by taking refuge in southern Canada and spreading further out in warmer times. These paleo-Eskimos lived in small tribes – the entire population probably never exceeded 3,000 – and despite sharing territory with highly sophisticated indigenous Amerindian populations in southern Canada, through social norms, they deliberately isolated themselves culturally and genetically. Paleo-Eskimo DNA has not been found in Indigenous American populations. Over time, the Eskimos struggled. Their health may have been poorer because of inbreeding, and their culture had evolved to a simplified form. They had lost social and technological complexity. They had regressed. By the time a new wave of Eskimos made the journey from Siberia – the so-called Thule people, or neo-Eskimos – the contrast between cultures could not have been more stark, even though genetically these were the same people. The whale-hunting Thules lived in large, well-organized villages and boasted advanced technologies such as dog sleds and sinew-backed bows. The paleo-Eskimos, on the other hand, lived in small villages of 20 to 30 people and hunted with chipped stone blades. There is no evidence of conflict between the two groups, but the paleo-Eskimos went extinct soon after; perhaps they were out-competed on resources, pushed to the fringes of the Arctic or simply annihilated by disease. An entire people disappeared through lack of trade with others.

Geographically isolated people are more vulnerable, but just one network connection can provide a cultural lifeline. When a much more recent Inuit population, this time living in remote polar Greenland in the 1820s, suffered an epidemic that selectively killed off many of its oldest and most knowledgeable hunters, the group lost its ability to make their most crucial and complex tools. Without specialist fishing spears (leisters), bows and arrows, or knowledge for igloos and kayaks, they were effectively marooned and unable to hunt their most useful foods. The population declined until 1862, when they were saved by another Inuit group from Baffin Island, who came across them on a hunting sortie and taught them essential cultural knowledge. The Greenland Inuit copied all the Baffin techniques, recovering their hunting and travelling ability, and from then on they made the broad, Baffin Island-style kayaks they'd newly learned. It took several decades, and an increase in population and reconnecting with other Greenland Inuit, for their kayak style to begin morphing back toward the sleeker kayaks of Greenland.

The idea that cultural evolution does not always result in progression may seem odd, but the same occurs with biological evolution. Darwin observed that, for example, some barnacles had genetically evolved simplified forms, even if the majority had evolved complexity. For human cultural evolution, population size and connectedness is key. Surveys by anthropologists show that larger populations have a greater number of and more complex technologies. One study compared individual Pacific islands with their population size and connectedness against the number and sophistication of their fishing tools:[41] Malekula Island, with around 1,000 people, has 12 different fishing tools; Hawaii, home to more than 1 million connected people, has more than 70 separate, sophisticated fishing tools.

Around the world, the societies that have survived have been the ones with enough genetic diversity to ensure health, and with a large enough social network to ensure cultural learning tends toward complexity.[42] Bigger groups mean a bigger collective brain – more minds to randomly generate successful innovations. Consider the invention of using feathers to fletch an arrow, for example,[43] and suppose that one individual operating alone will figure out arrow fletching once in 1,000 lifetimes. Then, the chance that one person in a group of ten will figure it out in their lifetimes is 1 per cent – a group of ten will take 100 generations to figure out fletching (2,500 years). In a group of 1,000 people, there's a 63 per cent chance they'll get it in one generation, and on average it'll take 40 years; for 10,000 people, someone will figure it out within a generation. More important for human cultural learning, though, is that bigger populations mean more teachers. Joseph Henrich designed an experiment in which students either had access to five different teachers or just one, to learn either image editing or knot tying and then pass on their skills to the next participant, who would in turn pass it on. For both tasks, the students who learned from five teachers improved their skills over the ten laboratory generations, whereas those who had access to only one teacher lost skills over that time.[44] Other scientists found similar results looking just at group size alone, determining that small groups are unable to maintain the ability to complete a complex task or improve on a simple one, whereas larger groups can improve both types of task over time.[45, 46]

Environmental factors that affect resource availability, and group mobility – how easily people can relocate to better areas – are also important for cultural complexity. Periods of drought and poor harvests, volcanic eruptions and tsunamis have all led to cultural disruption, population declines, and, in some cases, dark ages. But they have also produced the societal transformations that have accelerated cultural evolution through new interactions, migrations and technological transfers. Because we are a species that copies, rather than innovates from scratch, we can restore our cultural complexity

relatively quickly if it is lost, so long as we have access to technologically advanced people (just as the Greenland Inuits did). And cultures
can rapidly benefit from generations of cultural learning by other
tribes, by exchanging technologies. Consider that indigenous American 'Plains Indians' embraced equine technology almost overnight,
transforming their buffalo-hunting culture.

Trade networks, and their transmission of resources, genes and culture, were all affected by transport technology. The Yamnaya were
successful because they had horses and wagons, just as Columbus had
sailing ships. When the Romans built a network of roads across their
empire, they immediately boosted trade and innovation, and the effects
can be seen 2,000 years later: towns built on Roman roads are still
richer and more technologically sophisticated today.[47] Cultural diversity and complexity have always been most intense at trading posts.

As trade increased through larger, more complex networks and societies, collectibles like gold, silk, and shells became indispensable. They
were used to record debts, and as businesses grew, people needed to
borrow, increasing their debt. In North America, indigenous people
used shell beads called wampum, strung onto necklaces, as currency.
When the Dutch colonized New England, they embraced shell currency,
taking out a large loan from an Anglo-American bank in wampum.
Between 1637 and 1661, wampum became legal tender in New England
and trade flourished. European merchants also used their unequal
access to shells to manipulate local markets, for instance by flooding
Benin with billions of cowrie shells in exchange for industrial quantities
of slave labour. Over time, shell currencies[48] became standardized by
societies living in the Middle East and Europe, to the point where standard size was valued over beauty – this was the same cultural step that
gave birth to coins.

For complex economies with international trading networks, paying
in collectibles had its problems. Many states used gold or other precious
metals to pay for goods and services, weighing out the amount that cor-
responded to the value. People had to carry around lumps of actual gold
and silver and have bits knocked off them. This inconvenience was
made worse by the problem of purity – gold is often mixed with silver
and other metals in its natural state, and could easily be deliberately

tampered with. Archimedes famously solved this purity problem by working out its density, but this was time-consuming and annoying if you simply wanted to make a transaction. The solution was minted coinage – the state took away the purity headache by certifying and guaranteeing the value of the coins they issued: trade sped and eased. The first coins were invented around the same time (2,500 years ago) in Turkey and China, and were an overnight success, bringing great wealth. King Croesus of Lydia, Turkey, set the world's first gold standard when alchemists managed the difficult task of separating silver from gold and stamping the weight using the emblem of a lion. Coinage soon became the most important mass-produced item in daily use and transformed the economies that used it.

The next step in money was truly revolutionary: it was to extend the trust of the gold standard mark that people put in coinage to something that was inherently worthless. Paper money asked people to make a giant leap of faith and accept that although paper materially was of no value and was not in itself beautiful, the mark of the state's treasury gave it a value equivalent to its mark in gold. This requires trust on a national scale not only in the value of the note, but also in the stability of the institutions to keep the worth of the note intact. The first paper money, made of mulberry bark, was issued by the Chinese in the seventh century, and although it spread quickly around the region, it didn't catch on in Europe for nearly a thousand years. One problem was its vulnerability to forgery, but a bigger problem was managing inflation. Faith in the paper note was maintained by the promise that it could be exchanged at any time for the same value in coinage – in China's case, this meant small brass coins with square holes, known as 'cash'. But in the fifteenth century, the Ming emperor produced too many paper notes – the result was that their value plummeted and inflation soared. In 1455, paper money was eliminated in China and wouldn't be reinstated for hundreds of years. It was too good a technology to be abandoned forever though, and it's impossible to imagine the modern economy without it.[49]

I have a beautiful glass bowl in which I keep foreign currencies left over after trips abroad. Once, I used to go through my hoard before travelling overseas, looking for useful change. I have French francs, Yugoslavian dinars, Ecuadorian sucres, West German deutsche marks, and so on. But in my lifetime, most of these coins and notes have become archaic souvenirs, useless and of no value. Currencies have disappeared and been replaced as countries have broken up or changed. The main reason I haven't dipped into my dusty bowl in the past 15 years, though, is that in most of the Western world, coins and banknotes have been overtaken by new technologies that entirely remove the nationality and materiality from money. Credit cards, electronic transfers and cryptocurrencies enable international transfers at the touch of a button or swipe of a card. Trade is no longer done through face-to-face exchanges of beautiful collectibles, with the communication and reputational signals that entails. Now, we buy things from giant, faceless, global corporations and strangers with online addresses.

In 1996, Pierre Omidyar, growing tired of settling disputes between buyers and sellers on his AuctionWeb site, introduced a public feedback ratings system – users could award +1, –1, or 0 (neutral), and leave comments. The online reputational system was an instant success – eBay, as AuctionWeb is now called, makes more than $2 billion in revenue a year – and the technique has been used by traders in every type of enterprise to build trust between strangers. Electronic money transfers and credit card transactions are traceable and backed by guarantees and insurance, which helps grease the cogs of business between strangers in different countries and cultures. We can now buy almost anything from anyone across the globe and yet we rely, as our ancient ancestors did, on reputation and the invented value of collectibles for each exchange.

Trade with people outside of our social group of family and friends began with a decision to cooperate rather than fight. It started with exchange through barter and became about exchanging debt through tangible collectibles. But we have outsourced much of the cognitive accounting – the remembering of who owes what to whom, of who we can trust to do business with – to social institutions. We trade far more easily now, but the currency we use for transactions is no longer intrinsically valuable and requires a new sort of collective belief. However, the original contract we made with collectibles, to value biologically useless materials that cannot feed us, was perhaps our biggest leap of faith.

11
Builders

In 1965, in the village of Mezhirich in Ukraine, near the confluence of two rivers, a farmer was extending his cellar when his spade hit a hard, unyielding block. On inspection, the obstacle turned out to be the massive jawbone of a mammoth. He tried to dig it up, only to discover that it was interlocked with another buried mammoth jawbone. At this stage, the beleaguered farmer called in expert assistance. Excavations of the site revealed around 150 mammoth bones in strategic, interlocked clusters. The bones had once formed the frameworks of four spectacular houses, built some 20,000 years ago, when wood was scarce and the shelter of caves unavailable.

They were built by a hardy society braving some of the toughest conditions humans have endured, including forbidding ice sheets, blizzards and violent storms. The hunter-gatherers surviving in this frigid northern landscape managed to create grand, lasting structures of great beauty: the earliest evidence of monumental architecture.

These were remarkable, intricate constructions that took skilful planning and engineering to build. Each circular house began with a complete ring of inverted interlocking mammoth jaws, forming a solid base of roughly four metres across. About three dozen huge, curving mammoth tusks were used as arching supports for the roof and for the porch, some of them still left in their sockets in the skulls. Separate lengths of tusks were then linked in laces by a hollow sleeve of ivory that fitted over the join. Once completed, this extremely solid framework would have been covered with hides in a similar fashion to the skin-and-whalebone huts built by Siberian coastal hunters until the nineteenth century.

Each house required an entire herd of mammoth bones to

construct, although they were not all hunted – some of the bones had been scavenged, judging by the gnawing marks of carnivores. Nevertheless, the task of dragging enormous skulls, weighing at least 100 kilos, across any distance would have required considerable organization and cooperation. These buildings were clearly important to the society that had invested so much time, labour, resources and planning to build them. Mammoth bone was in itself a valuable material, and not just for its size: there is evidence[1] that it was of collectible importance just as elephant ivory is now.

Inside the bone buildings, there were beautiful treasures: amber ornaments and fossil shells, transported as far as 500 kilometres from their source, and the remains of one of the earliest percussion instruments ever found. The ochre-decorated drum, made of a mammoth skull set with animal long bones for drumsticks, show in their wear patterns how they were used, perhaps during rituals or other social occasions. Excavators also found the world's oldest map, engraved in mammoth tusk. Depicting the site from above, the map shows not only the houses, but also their position relative to the river and what might be a forest in the background. This place had meaning for its inhabitants – out of the wilderness they had created a home.

The primary purpose of the mammoth-bone dwellings was presumably as shelter from extreme cold and high winds. They are a cultural adaptation that enabled a tropically evolved ape to survive an arctic environment. They had to be built cooperatively but each one could fit as many as 100 people inside. Some archaeologists, impressed with the design, size and appearance of the structures, have argued that they also possessed religious or social significance. Similar bone houses have been found in considerable numbers, often clustered together in little 'villages' of four or five houses. They were made either by the same culture or by other tribes who had learned the technology. Further west, there was no need for such solid structures because limestone overhangs and cave mouths provided shelter.

We first used beauty to invent our personal and group identity, then to give value and meaning to material objects. But we also use beauty to design and define our environment. We invented monuments and habitats – ascribing spiritual meaning to naturally occurring structures, such as a mountain, or to cave dwellings, and then going on to create our own versions. We became builders, creators of symbolic architecture, homes and gardens. We began to make an artificial world somewhere in the space between our invented collective ideas and the physical landscape. Our buildings were designed and meaningful. We used the physical properties of the natural environment, reinterpreted them, and constructed a human world. In so doing, we would change the way we lived and operated as a part of the natural ecosystem. We would undertake a collaboration unlike any other species, drawing together dense networks of people to exchange genes, technologies, and behaviours, and become truly global.

The idea of building a 'home' goes back hundreds of thousands of years. Neanderthal constructions have been discovered deep inside Bruniquel Cave, in southwestern France, dating back 176,000 years. The structures, made from broken bits of stalagmite pieced together to make low circular walls, are among the earliest deliberate architecture known. The walls may have been internal separations to make cosy living quarters inside the cave, or they may have had a ceremonial or other purpose – there is evidence of fires being used there. In creating their homes, our own ancestors also modified natural architecture by decorating caves and rock shelters, and they made their own versions from the resources available. Archaeologists have found the remains of wooden partitions and lean-to structures that would have helped to exclude cold and damp, and evidence of cave-lion skins used as hut roofs.

We tend to think of hunter-gatherers as nomadic, which they are to a great extent, but most tribes have semi-permanent settlements, which may be in the same place for generations. Such camps may last for months and become the regular focus of feasting, religious ritual, and festival, as well as useful centres for trade. Many of the earliest camps would have been made of plant material, such as palm, wood, or bamboo, and long since deteriorated. Evidence of ancient semi-permanent open-air camps are being discovered in increasing numbers in western Europe, which appear to be summertime halts occupied by small groups of hunters who may have joined up with larger groups to live in the shelter of caves during the winter. At the best studied of these, a camp at Pincevent in the Seine valley some 15,000 years ago, researchers were able to trace the outlines of five reindeer-skin tents occupied by a small summer band of hunters. Although no foundations survived, the shape of the structures was revealed partly by the pattern of flint flakes struck off by toolmakers inside the tents.

Permanent structures like caves, some of which were occupied for tens of thousands of years, reveal the rich semi-settled lives of our ancestors. Astonishingly detailed artworks beautify these earliest homes, including Neanderthal paintings and stencils dating back 65,000 years, and figurative work made by people in Sulawesi more than 35,000 years ago, much of which was probably painted by women, judging from the size of the handprints. In making a habitat a home, humans increasingly transformed the environment – logging forests, hunting megafauna, and eventually creating entirely artificial landscapes. Our bodies mentally and physiologically respond to these home environments. We feel safe and comfortable in our home, and on entering, glucose tolerance, adrenalin levels, respiration and metabolism respond by changing measurably. This environmental stimulus triggers our biology in subtle ways, affecting everything from sleep patterns to fat deposition.

Among the stories our ancestors told around their campfires – tales that bound their listeners together into a cooperative group, building cohesive strength against the survival challenges they faced – were tales of supernatural beings. These myths, which often involved ancestors, gave spiritual power to iconic elements of the environment, such as the sky, rocks, lakes and hills. Animist cultures continue to worship and draw strength from significant landmarks, and many societies that have long embraced Christianity or Islam retain ancient beliefs in the power of a peculiarly shaped rock, a perfect cone-shaped volcano, or a beautiful animal, such as a jaguar. Once these symbols became collectively meaningful, people added their own decorative elements and involved them in rituals, such as painting Uluru (Ayers Rock) or taking on the costume of a jaguar.

Then people started building their own monuments, beginning the process of materially distinguishing humanity (and its cultural symbols) from the natural world. Some 12,000 years ago at Gobekli-Tepe ('Belly Hill'), in southeastern Turkey, a hunter-gatherer society made what may be the world's first megalithic building. Massive carved stones – some five metres high and topped with a vast horizontal oblong of stone – stand in circles on a hilltop. A few of these T-shaped pillars are blank, but others bear the evocative three-dimensional carvings of what must have been culturally significant animals, including vultures, foxes, lions and scorpions. This is decorative symbolism on a monumental scale – beauty not as a collectible to be owned and traded, but as a manufactured collective landmark to draw society together, and as a place to bury their dead.[2]

Now, the land is dust brown, barren and featureless, the result of millennia of intensive farming and recent climate change. But once, this was a fertile paradise, an auspicious spot for people following animals from the Levant and Africa. Here, at the time, would have been lush grasses, including wild barley and wheat; gentle flowing rivers attracting geese and migrating birds; fruit and nut trees; and grazing herds of wild herbivores.

These seven-ton pillars were not carved, erected and covered by an ad hoc band of wanderers. These were hunter-gatherers that cooperated on an unprecedented scale over centuries. The immensity of this symbolic, decorative undertaking would have required hundreds of people, who all needed to be fed and housed by the collective community. As it grew bigger and more famous, it drew more nomadic tribes to work on it, or simply in pilgrimage. Gobekli-Tepe would have been a destination for worshippers, traders and migrants seeking new opportunities. Settlements sprang up in the area and were used year-round to supply a growing population with food and other resources.

It was this urge to produce beauty – to make an enormous symbolic object to collective consciousness – that gave birth to the first permanent settlements over 10,000 years ago. Settled populations shifted our cultural evolution because they affected how we interacted as a society – the shape of our networks – and they also changed the dynamics of our interactions with the rest of the ecosystem.

Permanent human settlements put extra pressure on the local resources, meaning people exhausted the easiest foods and had to rely on eating less desirable options that needed greater preparation and were more costly to acquire. In order to feed so many people, these new villagers would have begun corralling wild sheep and goats, and planting concentrated gardens of wild grains and fruits, while weeding out unproductive, unappetizing plants. Just 20 miles from Gobekli-Tepe, scientists have found the earliest evidence of attempts at agriculture – a prehistoric village where radiocarbon analysis has revealed the world's oldest domestic strains of wheat, dated to 500 years after Gobekli-Tepe's construction.

People had been gathering and using the seeds from wild grasses for thousands of years, gradually shifting their evolution until they produced new species of domesticates. Chewed, fermented grains were brewed[3] at these camps, and this led to the evolutionary selection of genes that helped people digest alcohol. Brewing increased our knowledge of grains, some of which could be stored in the settlements. The idea of storing grains – itself a fairly revolutionary thought process – led to greater experimentation. People began selectively planting varieties with the fattest, easiest-to-husk seeds. Just as their ancestors had domesticated wolves to create dogs,[4] so they started taming grasses, changing their genetic evolution from varieties that had arisen through adaptive natural selection for wind scattering, to varieties that were adapted to human harvesting with sickles. Our new crops with their big, protein-rich seeds could be beaten and baked in the fire to make bread – a process that became so widespread and important in much of the world that we still see breaking bread with another person as an act laden with significance.

Our inbuilt urge for beauty – for visually expressing ourselves through meaningful material objects – had transformed us from tribal beings to trading tribes to settled farmers. At each stage, the human-carrying capacity of the environment increased – agriculture could produce around five times as many calories from the land as hunting and gathering. Bands of hunter-gatherers were small and moved regularly as they depleted an area's resources. However, through trade, and because of our strong social ties, our ancestors were able to significantly increase their population size – resources in one location could offset shortages in other areas. Permanent settlements soon became reliant on farming, which increased an area's carrying capacity still further so that, few as they were, the first Neolithic people out-populated[5] the hunter-gatherers everywhere they settled.[6] Farming was such a useful technology that it was independently invented several times across the globe, and widely transmitted elsewhere. Our built world depended on it.

Our ancestors may well have erected earlier monuments that have remained undiscovered or been destroyed over time. Around 70,000 years ago in Rhino Cave in Botswana, people sacrificed carefully made spearheads by burning or smashing them in front of a large rock panel, carved with hundreds of circular holes.[7] However, anything of the magnitude of Gobekli-Tepe, which required so much human labour over so long a period, could not have been made in conditions that couldn't support large populations. The earth's atmosphere during the last ice age contained so little carbon dioxide – perhaps the lowest ever seen – that photosynthesis became very inefficient, and consequently the planet's total vegetation was just half of what it is currently. Nomadic tribes living 20,000 years ago would not have been able to permanently settle in large numbers, because the spindly wild grasslands that managed to grow at that time, when atmospheric carbon dioxide was just 180 ppm (parts per million), couldn't have supported permanent herds, let alone farmers. Agriculture during the ice age would have been impossible,[8] and only agriculture can sustain large settled populations.

When the environmental conditions changed, around 11,000 years ago, with new ocean circulation patterns that replenished atmospheric carbon dioxide, Earth's ecosystem exploded in fecundity. Within 3,000 years, atmospheric carbon dioxide had risen to 250 ppm, leading to a phenomenal rise in plant productivity, which in turn helped invigorate the soils and store nitrogen and water. The vast increase in wild grains, fruits and other useful plants would have meant hunter-gatherers didn't need to travel so far for supplies, herds could settle for longer periods and people had enough stability in their resources that they could collaborate on huge monument-building projects. From such baby steps, we would become citizens and empire builders. Beauty transformed us and our world, but this cultural evolution was only made possible through environmental change.

Once humans made the cultural shift to agriculture and a settled lifestyle, this itself engendered further environmental evolution: we domesticated wild animals to make new livestock species, and wild plants to make crop species. By 5,000 years ago we had domesticated all the varieties of animal and plant we subsist on today – nothing has been domesticated since, and around 60 per cent of the calories that feed us come from just three grass seeds: wheat, maize and rice. This environmental-cultural evolution led to changes in our own genes, with selections for adaptations that help us metabolize cereals and resist the diseases of dense populations. We are genetically more different now from people living 5,000 years ago than they were different from Neanderthals. Positive selection in the past five millennia – just 150 generations – occurred at a rate 100 times higher than at any other period of human evolution, mostly because of changes in diet and disease epidemics, and because our larger populations have accelerated evolutionary effects. Some 7 per cent of our genes have experienced recent modifications.[9]

However, farming, especially early on, was an insecure lifestyle and many starved[10] or lived on the edge of survival – local wildlife would have been sparse as settled humans depleted it, and if harvests failed, migrating to new pastures was harder. Evidence from an archaeological site in Anatolia between 9,100 and 8,000 years ago, for example, shows that while there was a rapid expansion in population (mainly from a rise in birth rate), there was also an increase in bone infection and tooth decay from their starch-based, low-protein diets. The expansion of agriculture initially led to societal collapse.

It was not just health that took an agricultural hit: social well-being also changed, and many of these social injustices[11] persist today. The extraordinary settlement of Çatalhöyük – already a city, 8,000 years ago, of hundreds of one-room, mud-bricked homes, accessed from the roof – reveals evidence of a remarkably egalitarian society with strong social control and norms that prevented accumulation of wealth. By 6,500 years ago this seems to have changed, with a rise in inequality between households and a correlated rise in violent punishment for wayward members of society – skulls with deliberate attack marks that had healed have been found from around this time.[12]

This is also the time when the gender hierarchy emerged. One reason may be that the greater upper-body strength of men made them better able to plough, which meant that they got to dominate food supplies. Once men controlled such a vital resource, they had power over much else. In 1970, the Danish economist Ester Boserup[13] showed that the differences in the role of women in societies around the world correlate with the type of agricultural technology they use. Shifting cultivation, which uses handheld tools like the hoe and the digging stick, is labour-intensive, with women actively participating in farm work; whereas using a plough to prepare the soil is more capital-intensive and requires significant upper-body strength, grip strength and bursts of power to either pull the plough or control the animal that pulls it. Farming with the plough is also less compatible with childcare. As a result, men in societies characterized by plough agriculture tended to specialize in agricultural work outside the home, while women specialized in activities within the home. In time, this division of labour generates a norm that the 'natural' place for women is in the home. And this persists even if the economy moves out of agriculture, affecting the participation of women in all activities and employment outside the home. Many modern cultures in which the economy was previously based on hoeing or shifting agriculture, such as African ones, are more egalitarian than those that used to plough, such as Middle Eastern ones, research shows. A similar change occurred in sub-Saharan Africa, where the spread of cattle ownership led to a switch from matrilineal to patrilineal norms.[14] Matrilineal societies only persisted where the tsetse flies prohibited livestock farming. Environmental pressures can strongly influence culture.[15]

The type of agriculture also affects other social norms: rice growers, who use complex, multi-farm irrigation systems and depend on greater cooperation than wheat farmers, who rely on rainfall, tend to have a more collectivist mindset, a study in China found.[16] Wheat farmers, by contrast, tend to be more individualistic – more 'Western-minded'.

Whatever the technique used, as our ancestors increasingly settled to farm grain crops (which yield the most calories per area), this shifted social norms towards patriarchy. In a time when women's average life span may have been less than 28 years, and when 75 per cent of infants died, women were bearing and nursing babies all the time in order for the tribe to survive; although nomadic tribes space their infants, being unable to carry more than one at a time, agrarian societies had babies as frequently as every year. Men began to control women's reproductive capacity for economic reasons as child labour tilled the soil and shepherded the herds. Men began to control the sexuality of their wives, to be sure of paternity for children that they were supporting and, as they accumulated resources, to be sure of their heirs. Women also suffered from the practice of inter-tribal exchanges of women for marriage, which uprooted young women from family support and meant that it paid more for men to form alliances with male kin who would remain nearby for a lifetime. Settled agriculture led to the increasing domination of patriarchal warrior tribes[17] over egalitarian societies lacking a warrior class, resulting in women and children becoming prisoners and slaves, in an era when male prisoners were killed.[18] In short, women and children became the property of men.

Settled farming also had other enormous social repercussions. For a start, it relied on far greater cooperation with non-relatives on bigger public collaborations, such as protective battlements – once people were invested in an area of land with crops and other property, they needed to protect it against other tribes. Agriculture on the scale needed to support larger villages and cities also required large-scale earthworks, such as the digging of irrigation channels, protective dykes and channels. Such projects needed planning, organizing and managing with the establishment of hierarchical structures and institutions, which altered social networks, people's positions in them and thus their life chances.

Consider that hunter-gatherers obtain what they need from their surrounding environment as they need it. There's nothing heroic for them about generating a surplus; spending time getting more food than you can eat is a foolish waste. Settled farming, though, would generate a whole new economy – taxes – that then supported more public building and also the social infrastructure to increase population, in a mutually reinforcing cycle. Grains could easily be taxed per field area because the crops were harvested at a predictable time, could be stored and traded and even used in payment as a currency. A taxable populace allowed new social structures to form with an elite that exerted power over a state and used crop surpluses and taxes to fund measures, such as infrastructure, armies and city walls.

States didn't commonly form in places where the staple crops were tubers, like cassava, because avoiding tax is so much easier when you can't actually see the food and harvest times are so variable.[19] Agriculture was very labour intensive, and, once states relied on production levels and tax incomes, the labour itself became as important a resource as the cereal – elites managed these resources ruthlessly, striving to maintain large workforces (at a time when life expectancy was falling due to disease and malnutrition) by enslaving tribes through warfare and controlling peasants in some form of bonded labour. In some taxation models, the state used tariffs to help feed the poor, usually as a way to buy loyalty and prevent unrest. In ancient Rome, for example, populations of conquered provinces were taxed, and these funds helped provide the dole for poor Roman citizens, who were exempt from taxes.[20] There were vast numbers of unemployed citizens among the city's population of 2 million, and dangerous mobs were avoided via the policy of 'bread and circuses', in which citizens received free food and entertainment.

The considerable socioeconomic changes to people's lives with agriculture, in which people survive hardship and invest time and effort on big projects on the promise of eventual reward and harvest – long-delayed reciprocation – requires faith. Beauty plays a big role in this. Monuments are physical embodiments of hope and represent a grander power that flawed mortals can defer responsibility to. From this, we evolved the idea of a state – itself a conceptual monument, in which we embed values and meaning, and in which we align our individual and tribal identities.

Some of the biggest, most dramatic monuments have been made by the most hopeless people who, common sense suggests, would have been better off using their time and resources to feed themselves. But that is to underestimate the value of meaning – monuments give shape to our cooperative strength. Take the iconic statues of remote Rapa Nui (Easter Island), monumental ancestor figures called Moai that speak silently of a tragedy. More than 3,000 kilometres offshore, this was one of the last places on Earth to be permanently inhabited, some 1,300 years ago, when sophisticated Polynesians, able to 'read' the oceanscape (like the Kalahari Bushmen read the savannah), arrived on double-hulled canoes. By recognizing particular patterns in the waves, analysing the types of floating debris, cloud formations, weather and a host of other culturally learned skills, they were able to commute between trading posts from New Zealand to Fiji and beyond.

However, by the sixteenth century, the Rapa Nui population had lost this ocean-navigating knowledge and was in dire straits. Mounting environmental pressures had devastated agricultural yields and in desperation, the community responded by carving hundreds of giant stone Moai. Their shared belief in the protective Moai gave them strength, drew them together – a culturally evolved human survival adaptation. But the very act of making these decorative monoliths, which measured as much as 21 metres, and transporting them from the quarry, involved deforesting the island for rolling logs. This caused soil erosion and worsened the drought, eventually leading to a famine that severely reduced the population. Inter-tribal wars between clans broke out, with villagers toppling each other's Moai, killing rivals, and eating them. A presumably effective insult was 'the flesh of your mother sticks between my teeth.' Moai ancestor worship diminished in 1600 and was replaced by a Birdman cult – an exceptional example of a culture dramatically changing religion without outsider influence.[21] The Birdman cult was a spring-worshipping celebration, focused on the island's scarce environmental resources.[22] Cultural evolution had brought environmental change that led to further cultural evolution.[23]

Living in far bigger numbers requires new social institutions and the strengthening of others – reputation is still important, but underlying this are hierarchical structures and the creation of tribes within tribes, which form automatically once group size increases, and allow more people to be fed due to economies of scale. Eurasian agricultural societies became more unequal than North American ones, perhaps because the Eurasians had the energy advantage as they had large domesticated animals – horses and cattle – whose labour enabled more rapid and extensive economic growth – and competition for these resources would have exacerbated inequalities.[24] This hierarchical structure is woven into the social narrative that binds the tribe together and is strengthened with social norms and decorative iconography, making it increasingly hard to challenge authority, establishment figures and orthodoxy. Monuments and symbolic art reflect these social norms – wealthy people have been regularly portrayed as gods or closer to gods. They owned not simply more land and food, but also people's livelihoods. By logical extension, poor people become less godly, less good and feckless, deserving of their situation and beholden to the generosity of those wealthier.

In almost all societies, an individual's position on the social network continues to be defined by birth. India famously retains a strong caste system (so strict that inbreeding over generations is noticeable in the genomes);[25] in Britain too, the socioeconomic status of parents strongly predicts the child's future profession and wealth – when parents send their children to elite schools, what they are paying for is social selection, a central position in a network of elite peers, who will go on to make up the majority of business heads, political leaders and opinion setters. At the other end of the social scale, until recently, were France's own 'untouchable' caste, known as Cagots,[26] who for hundreds of years were treated as inferior and confined to dismal ghettoes known as Cagoteries.

Inequality is a problem in bigger societies because we have an innate sense of fairness,[27] and while worker bees do not, as far as we know, yearn to be drones or queens, humans desire beauty, meaning and happiness in their lives – even agency. There is continual tension between the desire for individual autonomy and the collective good. Over millennia, societies have grappled with the problem of how to keep large populations of socially unequal people from revolting. The Chinese philosopher Confucius aimed to create a fairer, happier society by tapping into our human search for meaning and individual self-expression. He proposed running society along the hierarchical lines of the family, with everyone in their place: the 'paternal' emperor ordained by the gods, with the rest of society controlled not by intimidation but through the value systems of (patriarchal) families, using mutual consent, honour, respect and love. People were taught to cultivate moral virtues in their everyday behaviours, which would then produce a good society. This kind of practical philosophy, in which an individual is assumed to have agency over their behaviour and their society, forms the basis of many of the world's great teachings, from Socrates to Jesus, and it's an attempt to make sense of an uncontrollable world by focusing on relationships between individuals. The collective message, which has proved enduringly popular, is that through empathy and kindness to another human we gain our own humanity.

The move to agrarian culture produced lasting environmental change. The transient relationship that hunter-gatherers had with the landscape gave way to permanent changes as settlers scooped clay from river-beds to build their houses, altered the drainage and course of rivers, deforested, extensively grazed and eroded the soils. Neolithic people pioneered large-scale environmental change, beginning the process of turning vast areas of natural woodland, marsh and grassland into the agricultural landscapes of artificial monocultures we recognize today. Farmed soils soon became depleted of essential nutrients, such as nitrates and phosphates, which were very hard to replace. The most effective way of replenishing soils was to burn down forests and other vegetation and grow crops on the rich ashes. Slash and burn agriculture, or swidden, soon converted Europe's wild landscapes, which were further fertilized with collected dung from people and livestock. At the same time, people also created a profusion of artificial 'caves' – big longhouses, capable of housing several families, sprung up across Europe by 8,000 years ago.

This was a quite fundamental shift in our ancestors' perception of the natural world and their relationship with it. Most hunter-gatherer societies consider themselves an integrated part of the ecosystem, and their culturally evolved behaviours and technologies reflect this, as we've seen, with social norms that limit hunting at certain times of the year or in certain places. These have probably evolved for pragmatic reasons to avoid unsustainable harvesting of the resources on which their survival depends. However, once people began to own animals and plants, as opposed to interacting with them as one wild creature to another, this relationship shifted. As we transferred our gods from living animals and natural structures to our constructed, monumental representations and to human forms, we also changed the hierarchical status. When we built permanent structures that sheltered us from the elements, paved our routes and redirected the flow of water, then we made a human world that was increasingly distinct from the natural world. We created an artificial environment that was blissfully free of the inconveniences of cold, wet, muddy, dangerous nature. Meanwhile, we chose to encourage those bits of nature we wanted: our own artificially modified species of foods, beasts of burden and material resources. A recent study[28] explored this transition, comparing how children growing up in an urban Chicago environment and children from a Native American Menominee society played with toy animals.

The original experiment[29] was adjusted to include a diorama with realistic trees, grass and rocks, after one of the Menominee elders complained that it made no sense for the children to play with the animals divorced from their ecological context. The study revealed that while the Chicago kids gave the toy animals human attributes during their play, the Menominee children played with the toys imagining they were the animals.

Once we built our own world, we began to see humans as separate and dominant over the rest of nature,[30] and we began to see nature as only valuable where it produced a useful resource. This would profoundly change our environment and the evolutionary trajectory of countless species.

We know that agrarian technology spread widely but, until recently, we've not known how – whether it was the idea that was traded, or whether it was the people themselves who migrated. Now genetic analysis is giving us a better idea, and it seems that a bit of both took place. Within the Fertile Crescent area, it looks like the idea of farming was traded between populations, along with tools and collectibles such as obsidian. Small populations of these farmers migrated from Anatolia – one group made it up into Europe, bringing their Linear Pottery, new seed-collecting and sowing technology, brewing expertise and animal husbandry with them to colder climes and less-conducive seasons; another, from the Levant, probably travelled to East Africa – one-third of Somali DNA comes from the Levant population.

From the DNA evidence,[31] we know that a population of these first Anatolian farmers travelled to Europe, around 9,000 to 7,000 years ago, and gradually began to partially mix with the hunter-gatherer locals. It was no trivial thing to transfer crops such as barley and rye to the northern fringes of Europe: these agricultural pioneers were taking crops that had evolved over millions of years in the Middle East, where they had adapted to a dry-wet pattern of seasonality, and moving them into an area that was recently de-glaciated. These skilled, experimental farmers were the people that created glorious monuments, such as the Stonehenge complex, using an army of builders fed on the first farmed produce and supplemented by wild food such as boar and auroch (the extinct ancestor of domestic cattle).

When the Yamnaya arrived in Europe, the agrarian and pastoralist cultures were unified by strong social norms of property and land ownership, and the new idea of transmitting property between related individuals and families. It is partly because of these norms that there are still genetic, if not visible, differences between groups in homogenous Europe – even within Britain. A fine-scale genetic map of DNA in British people[32] shows distinct clusters of those whose ancestors have lived and married in the same small geographical area for generations, and others who can trace their ancestry to successive waves of immigrants. People in the Orkneys are, unsurprisingly, the most genetically distinct, with strong Norwegian Viking heritage. Elsewhere, though, it is quite remarkable to see genetic distinctions across seemingly arbitrary borders, such as between the people of Devon and Cornwall, where the county boundary along the River Tamar (established in 936 by King Athelstan) has genetically separated populations for centuries. Elsewhere, those in North Wales can trace their ancestry to perhaps the first British inhabitants, but their Celtic cultural links to Irish or Scottish populations aren't reflected in their genes. Sometimes cultural practices are traded between people or imposed upon them; sometimes the people themselves migrate and integrate, bringing their practices with them. The new advances in population genetics, combined with ancient DNA, archaeology, paleontology and linguistics, are starting to give us a much fuller picture of how our cultural evolution occurred. Anglo-Saxon migrants came to Britain and we can see where they settled down to individual towns because they changed the gene pool there; the Romans, Vikings and Normans, whose invasions transformed British cultural history, left no legacy in our living DNA records.[33]

Population genetics across Europe reveals a similar tale. Geneticists carrying out one study on 3,000 people, noted:[34] 'A geographical map of Europe arises naturally as an efficient two-dimensional summary of genetic variation in Europeans.' It's worth remembering, though, that these tiny clusters of genetic similarity or differences between populations, whether between the Cornish and Devonians or Sri Lankans and Swedish, are, whether invisible or highly visible, just small genetic anomalies in an otherwise very homogenous species. All living humans are now pretty similar – much more similar than two chimpanzees – and that's because we emerged little more than 200,000 years ago as a relatively small population, and we have since experienced substantial population crashes (genetic bottlenecks) and a lot of migratory breeding, facilitated through trade networks. Any two humans now differ by an average of one in 1,000 DNA base pairs (0.1 per cent), which shows remarkably low genetic diversity compared to great apes.[35]

If you divide humans by continent, approximately 90 per cent of genetic variation between individuals can be found within these populations, and only 10 per cent between them. This is, in part, because we are all related. I don't mean related in our deep ancestral past, but relatively recently. You don't have to go back very far in your family tree to find a common ancestor. Consider that I have two parents, four grandparents, eight great-grandparents, sixteen great-great-grandparents, and so on. Go back like that 40 generations, about 1,000 years, and each of us would have around a trillion ancestors, which is far more than the total number of people who have ever lived. It only makes sense once you factor in the overlaps – or interbreeding – which means, as you go back through the generations, your relations play more and more roles in your ancestry. Thus, your great-great aunt may also be a great-great cousin, and possibly also your partner's great-great cousin. Statistician Joseph Chang worked out that all of our family trees are interlinked with each other's within a few generations – a sort of six degrees of separation back through time.

Anyone with European ancestry descends from the Emperor Charlemagne. In fact, everyone who was alive in Europe 1,000 years ago and has living descendants – that's 80 per cent of them – is the ancestor of everyone alive today. And you only have to go back 3,000 years to find the most recent common ancestor of everyone alive on Earth today.[36] So, I am also descended from Muhammad[37] and, in common with pretty much everyone else on the planet, I'm descended from Confucius and Nefertiti. This also means that if my kids go on to have children of their own, in a few thousand years time, I will be the ancestor of everyone alive on Earth.

The dense interconnectedness of the human family, our genetic similarity, makes us mongrels and means there are no different races of people.[38] Genetic differences do exist between populations, but their impacts are rarely behaviourally or biologically significant in comparison to the effects of culture. Usually, the combination of environmental, cultural and genetic conditions that led to selection for a variant will also factor into how it is expressed. Pacific Islanders, whose ancestors took long ocean voyages with little prospect of food, evolved a genetically adapted metabolism in response to that cultural pressure. Now, though, their cultural developing bath has changed. They lead a sedentary lifestyle, eating high-calorie imported food, and this, combined with their genetic variants, has made them the fattest people in the world, with alarming levels of diabetes. There are multiple genes involved in metabolism, and lifestyle factors play a huge role in obesity – the problems in Fiji and Polynesia are overwhelmingly cultural, even if genetics plays a small role.

Even for a trait like height, which is 80 per cent heritable, there are big differences between people living in poor countries and those with good nutrition. People who grew up during war and famine are often shorter than their well-fed children. The Dutch have increased in height by 20 centimetres over the past two centuries of economic growth. Indian girls and non-firstborn boys are on average stunted because a cultural preference means the firstborn boy gets the best nutrition and grows tallest.

However, in one South Pacific island, the effect of genetics is overwhelming. In 1780, a volcano all but wiped out the population of Pingelap; as few as 20 people survived. The island's relative isolation, combined with a social norm that discourages marriages to outsiders, allowed genetic mutations to accumulate in the population. Now, as a result of this inbreeding, 10 per cent of the island's population is severely colourblind, meaning they see only in black and white. The mutation is a problem in daylight, but at night, its carriers see better than those with normal vision, making them excellent at night fishing, and pointing to an explanation for the gene's persistence.

It had been assumed that once people developed agriculture, the tribal differences in genes and culture that had developed in small, isolated hunter-gatherer populations broke down because people lived in expanding villages and moved between them. Unusually, though, Papua New Guinea is still genetically very diverse despite having agriculture. It seems that what made Europe, East Asia, and sub-Saharan

Africa different was that they had a Bronze Age and a subsequent Iron Age. It was the trade networks arising from this technologically driven expansion that changed the culture of these regions as people started travelling to trade, and over time this produced a more genetically homogenous region. Papua New Guinea, by contrast, still reflects the levels of genetic and linguistic diversity that European societies would have had before the Indo-European-speaking Yamnaya arrived with their metalworking culture. Now, only the Basque language survives as a remnant of Europe's hunter-gatherer past.

Geography and environment had a big influence on how human populations intermixed, traded and transmitted culture. The Eurasian landmass is wide rather than tall, meaning that the same agriculture can be practised for thousands of kilometres from east to west because similar latitudes have similar climates. When Eurasians arrived in North America, they could grow the same crops and keep the same livestock, and the same was true for South Africa and Australia. North and south, however, in Africa and Latin America into the tropics, agricultural practices and technologies needed to adapt. Logistically, Europe benefits from an excellent network of waterways that helped transmit culture much more readily than in Africa and South America, with their unnavigable rivers, mountains and other transport blockages.

A more subtle but important biological barrier to population mixing has been disease resistance, which is partly genetic and strongly influenced by environmental factors. Once we started living in dense, large populations with the advent of settled agriculture, regular epidemics of diseases that were contracted through close contact with other people or animals spread through communities – those that survived passed their resistant genes on. Plagues sweeping through Europe and Asia changed the course of history, bringing down empires and ushering in new waves of people and cultures. One intriguing legacy from these devastating plague and smallpox epidemics is that the European descendants of survivors may in their genes carry resistance to HIV.[39] Another consequence of Europe's lengthy exposure to infection was the rapid conquest of Australia and the Americas, after the indigenous populations succumbed to smallpox, measles and flu, consequently transforming the geopolitical and cultural map of the world.

Meanwhile, Europeans trying to conquer African and Asian rain-forest territory in search of collectibles like gold, diamonds and ivory were beaten by native diseases, such as malaria. The indigenous resistance there is coupled with a high incidence of hereditary sickle-cell anaemia, which offers malaria protection (because the parasite can't live in the oddly shaped haemoglobin), but is weakening because the blood can't carry enough oxygen. Yam cultivation creates a perfect mosquito-breeding environment, so African populations with a history of yam cultivation have a higher incidence of sickle-cell anaemia, but a lower mortality from malaria. As we change our environment, so we change our genes.

Genetic differences between populations are fading, not because we've stopped genetically evolving, but because we are intermixing more. Much of the tribal isolation of the past was ended by sexual dalliances between groups, intermarriage, migration and trade. Even when tribes held strict norms forbidding intermarriage between groups, the genetic evidence reveals it continued. The domestication of the horse and the invention of wheeled transport accelerated this, but well into the nineteenth century, Europeans were still marrying close relatives. The bicycle[40] reduced this considerably, by enabling sex between geographically distant populations. The sale of 4 million bicycles before the First World War had a striking impact on French society, including making the French taller by reducing the number of marriages between blood relations.[41] The same effect was seen in England.

Humanity's ultimate monument is the city: an entirely artificial landscape, designed and built with intention to symbolize the culture and aspirations of its people. This is planetary-defining beauty – a resurfacing of Earth's skyline that can be seen from space. Our cities are constructed for beauty and meaning, often at the expense of their functionality, and become almost a living embodiment of their citizens. When Paris's Notre Dame Cathedral was consumed by fire in April 2019, it elicited an immediate and strong emotional response from people around the world. The tragedy was not that local Christians would have nowhere to worship, or that spacious shelter had been lost, or even the economic hit to tourism revenue. The heartbroken citizens reacting to the blaze were mourning an integral part of

themselves – we are each the result of our genetic and cultural heritage, and the centuries-old monument is a visible part of that anatomy.
Within days, hundreds of thousands of euros had been pledged for its
restoration.

In making cities our habitat, we accelerated our cultural evolution.
Just as the Silk Road and the Atlantic Ocean became vital networks
for the exchange of ideas, technologies and genes, so cities too
became focal points for such cross-cultural trade. Cities are cultural
factories because they attract people from diverse populations into a
dense environment that increases opportunities for them to interact.
As the trade networks grew and technologies evolved, cities could
become ever denser, increasing the rate of innovation in a positive
feedback loop. By the thirteenth century,[42] Londoners relearned from
European merchants the art of timber framing (lost after the Romans
left), allowing multi-storey buildings to be erected and facilitating a
dramatic rise in population density. By the end of that century,
Cheapside had townhouses with three storeys and a garret.

Cities, like all social networks, are synergic: the sum of their impact
is greater than the parts. Increase the population of a city by 100 per
cent and creativity increases by 115 per cent.[43] Cities cannot exist in
isolation: they rely on trade networks of merchants, diplomats and
artisans, who bring new resources and ideas from other places. Ideas
are incubated in the streets, coffeehouses, universities and institutions of cities, and evolve into the diversity of technologies, arts, and
cultural practices we see today. The genetic effects of this are so powerful they can be seen centuries later in their living descendants. Some
400 years ago, a charismatic leader of West Africa's Kuba tribe, called
Shyaam a-Mbul, formed a peaceful kingdom (in what is now the
central-southwest of the Democratic Republic of Congo) that drew
together the region's many ethnic tribes into a sophisticated, large
city-state. The Kuba Kingdom – with its incredibly modern political
system, including a constitution, elected political offices, trial by jury,
public goods provisions and social support – became a very wealthy
hub of innovation, famous for its artworks. When the first Europeans
arrived there at the end of the nineteenth century, they wouldn't
believe such a familiar political system had arisen independently and
assumed there had been earlier contact. Belgian colonization did

much to diminish this remarkable cosmopolitan state, yet its legacy is defiant, living on in the DNA of the descendants; populations with Kuba ancestry are far more genetically diverse and carry signatures of a wide range of ethnicities, compared with others from the area.[44]

Citizens, because of their relative anonymity (which weakens the reputational pressures to follow norms) and the strength of their smaller tribes, are more able to invent new social norms, from gender differences to musical fashions. Decoration is key to reinvention. Take the ubiquitous clay tile, used for millennia to cover floors, walls or roofs. The extraordinary range of decorative expression, from domestic and pastoral scenes to religious narratives, reveal their societies' changing ideas over time. And because these decorative norms embody social norms, they embody (and build) group identity. In 668, the new unified state of Korea demonstrated its wealth and power with a massive construction program. The capital, Kyongju, was to be a splendid planned city of 180,000 new houses, roofed with expensive tiles rather than thatch to make them weather and fireproof. Each ridge ended with specially decorated dragon-motif tiles, which became emblematic of the new state's strength and is still used today. The humble tile had become a monument to statehood.

Cities represent the greatest expression of our urge to produce beauty and to triumph over the natural environment. Enormous effort is spent on beautifying the spaces we inhabit and conveying meaning through architecture. From the great Ziggurat of Ur to the Harpa concert hall of Reykjavik, we have created impressive monuments with precious materials and human hours, distilling our hard-won survival adaptations into decorative structures that last beyond our bodies, and potentially our genes.

Cities are built environments that selectively evolved under cultural pressures. And these new environments we have made are, in turn, changing our biology and our cultural evolution, as well as the genetic evolution of the natural world. Birds have adapted to urban environments with louder songs, longer beaks adapted to feeders and plumage changes. Cave moths introduced to Europe in the past two centuries are now clothes moths, wholly dependent on urban home furnishings to survive. We humans are also dramatically affected by urban living. Diseases swept through malnourished, packed communities, and the infrastructure itself added to the problem, piping in disease and toxins, including lead[45] in ancient Rome and in modern-day Flint, Michigan. Urban air pollution today causes cardiovascular and respiratory problems, and is responsible for 9 million deaths a year. It's worth noting that while cultural evolution has produced fancier technologies and social institutions, these haven't necessarily improved most people's lives or life spans, even if population increases. The great cultural advances of the Roman Empire were actually pretty disastrous for the health of its subjects. The average femur length for men in Britain shrank over the course of Roman occupation, then rebounded after they had left: 'The Romans were helplessly caught in the vice grip of their own progress, with its confounding ecological repercussions.'[46] Urbanism, with its greater concentrations of people living in unsanitary conditions, was much of the problem, but also new imperial networks transmitted diseases. Archaeologists can now track the spread of the empire by tracing the spread of intestinal worms.

Sanitation has always been an issue for densely packed human populations, and city living was, until very recently, significantly life-shortening. Such was the death rate that their populations were only maintained through the regular immigration of countryfolk. Life expectancy in 1861 of a boy born in Liverpool was 26, compared to 56 for one born in Okehampton in Devon. Keeping clean was best achieved, it was believed, by wearing a linen shirt that could be washed, for washing your actual body was thought to risk plague and other deadly consequences, and was thus strenuously avoided by Europeans[47] for well over five centuries, until the late 1800s. With the notion of germ theory and the public investment in sanitation that followed cholera epidemics and events such as the Great Stink of London[48] in the summer of 1858, it became easier – and desirable – to keep our bodies clean. Social norms changed and attractiveness became linked to how clean we and our clothes were. Cleanliness became an important and achievable goal, leading to bathhouses, toilets, sewerage and a whole industry devoted to perfuming away the natural smells of a crowded humanity.

The cultural switch from hunter-gatherers to agriculturally dependent urbanites increased stratification of society and, while being of immediate benefit to a small elite, worsened the diet and health of the great majority, and had enormous impacts on ecosystems. However, evolution is rarely simple. Trade that brought wealth and ideas to western Europe also brought the Black Death epidemic that slashed populations, and this produced its own environmental change. As the population fell, so did human agricultural activity – forestry increased and pollution decreased, causing a measurable regional cooling. (The same environmental consequence occurred in the Americas when indigenous farmers were devastated by European epidemics.) In England and Wales, the drop in food production because of plague deaths led to dramatic social and agricultural changes, as formerly open common lands were enclosed and peasants gained more rights and ownership over plots of land, which incentivized innovation and investment. Whereas before, agricultural land was left fallow to allow soils to recover while livestock grazed, newly enclosed land was intensively farmed using a crop-rotation system in which shallow-rooted grasses like wheat were followed by deep-rooted tubers like turnips, then legumes like clover that restore nitrogen to the soils. Previously, there had been little incentive to plant root vegetables that risked being dug up by other people's livestock on common land. Technologies such as the adjustable Dutch wheel-less plough (actually a Chinese invention) allowed boggy land to be worked using just one or two oxen rather than six or eight, and enabled moors and fens to be drained. Agricultural production soared, becoming the highest in the world, and the surplus food could be traded in expanded networks. Populations rose, and it was on the back of this new workforce that the industrial revolution and our modern world was forged.

We are still biologically evolving to live in urban environments – humans find them stressful, and this is linked to a rise in psychiatric conditions[49] such as schizophrenia and psychosis, as well as behavioural problems[50] and autoimmune conditions such as asthma. And the urban environment may also be delivering epigenetic changes – these are changes to the way a gene is expressed. Pregnant women living in stressful, polluted cities are more likely to have babies with negatively affected brains, metabolism and immunity,[51] and these genetic changes may be passed on to the next generation[52] – the interplay of our cultural-genetic-environmental human evolutionary triad in action. However, despite the health risks, cities are hugely seductive.

They represent the enlargement of our tribe, with the associated benefits of monetary and cultural wealth.

The Internet, which acts as a virtual city, may be producing similar cultural effects as it broadens the social connections that people make. Steve Jobs once described computers as bicycles for our minds. More and more of us are now hitching up with complete strangers we first discover on the Internet. A mathematical model of online connections predicted that these networks should result in markedly higher rates of inter-racial marriage and a lower rate of divorce (because of the greater compatibility of matches). The number of inter-racial marriages has already shot up in the United States since the introduction of online dating sites. Great migrations of people, invading, fleeing, crusading, exploring, roaming, colonizing, slave trading, uprooted for war, work, or fortune, and now the Internet – all have contributed to the great genetic mixing of the past millennium, particularly in recent centuries. This is leading to a few hiccups, such as a rise in vitamin D deficiency in dark-skinned people living in northern latitudes. But we are heading towards a situation where tribal in-group/out-group prejudice and attractions will no longer be possible based on visible distinctions. In other words, describing differences between people based on the fallacy of 'race' will no longer be credible.

While all animals are driven by biological urges to find food and mate, humans are also motivated by meaning and purpose. We find this in beauty; we also find it in the quest for knowledge, as we shall explore next.

TIME

How do we know what we know? We, who are the inheritors, the cultural and biological legacies of our ancestors, question our existence; we wonder about who we are and where we are in space and time. We have our stories, and they tell us about the past and help us imagine a future. But we are haunted by the idea of reality, an objective truth, and we are driven to pursue it. We have spent our entire existence trying to grasp the intangibility of time, to mark and even control it. We observe, predict, measure and reason out our mysteries to decipher the future, and in so doing we have re-created the world and ourselves within it.

12

Timekeepers

In 1962, a young French geologist holed himself up in a cave deep inside the Alps for two months. Michel Siffre wanted to see whether our bodies need outside stimuli, such as sunlight, to maintain their natural rhythms or whether there was some sort of timepiece inside of us. 'I decided to live like an animal, without a watch, without knowing the time,' he said.

Siffre's experiment was a gruelling endurance test. The cave he selected was an icy cavity, accessed via a dangerous and technically difficult descent along a 45-metre-long, S-shaped shaft. Just getting there with all his equipment was fraught with danger; rescuing him, in case of injury, would be near impossible. And yet the 23-year-old insisted on his isolation during the experiment: for the first month, his instructions were that he was not to be rescued, whatever the circumstances. During the long nine weeks, he kept meticulous records of his physiological measurements, everything he ate and his state of mind. Two researchers were stationed at all times in a tent up at the cave's mountain entrance, with a telephone line to Siffre's cell. Whenever he woke, Siffre would phone them and they would answer and record the time of the call.

With no idea whether it was day or night, his body nevertheless soon adjusted to a sleep schedule, and the young scientist, growing increasingly miserable and uncomfortable in his damp, freezing home, tried also to mentally adjust to the new environment. 'I had bad equipment, my feet were always wet, and my body temperature got as low as 34°C. I spent a lot of time thinking about my future,' he recalled later.[1]

In his lonely, uncomfortable state, Siffre's appetite diminished and his diet shrank to bread and cheese, while the two records he brought *with him soon failed to interest him. The only pleasure he found was in a spider he captured for a pet. And what of his timekeeping? By the second morning, Siffre-time was already two hours off real time; by ten days, he thought day was night and noted in his diary how the researchers' cheery 'hellos' revealed they had been up for hours – in fact, he had regularly woken them in the middle of the night. During his phone calls, Siffre took his pulse, counting from one to 120 over two minutes. However, his colleagues above ground could see that Siffre's two minutes lasted five minutes.*

He endured the solitary confinement, rationing out his favourite cheese to last the distance; then suddenly, with 24 days of the experiment to go, he was told it was over. The researchers announced that the time was up and they were coming to get him. Siffre's own estimation of how he spent his time in that cave were so out of sync with actual time that he had managed to 'lose' around one-third of his 63 days. Some of his seemingly brief 10- to 15-minute catnaps had in reality been eight-hour slumbers. Without any way of telling day from night, time had slowed for him. But it hadn't slowed for his body. Despite his confusion, his DNA had kept his body ticking to the same timetable below as above the ground.

We are all creatures of time. We evolved in a universe of interwoven space and time, and our bodies are adapted to our planet's movements. All of our cells have clock genes, which interact with each other like the cogwheels of a mechanical watch, generating oscillations of gene expression. These timepieces regulate our genes, hormones, heart rate, brain activity, moods and bodily functions.[2] Our bowels are most active at around 10 AM, whereas we have the highest tolerance to pain and the best coordination at 2 PM. At 5 PM, we are at our physical peak, with maximum muscle strength and flexibility and the best lung and heart performance. Our alcohol tolerance is best at 8 PM, sleep hormones begin rising an hour later, and we are in our deepest sleep between 2 and 3 AM.[3] Our body temperature drops to its lowest between 4 and 5 AM. Our bodies obey their biological timetable with remarkable regularity from menstruation to gestation.

While our bodies evolved time-telling, our conscious minds did not – and human culture depends on conscious decision making. The passage of time, the cyclical solar system in which we are immersed, affects all aspects of our culture, and so humans had to evolve the cognitive tools to time travel, and the cultural tools to consciously track time. It is only through mastering time that we could create complex, sequence-dependent technologies, hierarchical social structures and language (for which word order and sentence structure define the meaning). Time, though, is an abstract, invented notion that our ancestors learned to collectively believe in and to manipulate until we became superlative mental time travellers, able to relive past episodes (even ones we never experienced), and to project ourselves forward to an imagined future.

As far as we know, we are the only animals to understand that sex leads to the birth of babies; we are able to understand that an activity today will have consequences nine months later, and we can also, therefore, trace our kinship, thus broadening our networks. Humans also know their mortality – that we will one day die. It is perhaps this ability to 'sense' time passing irretrievably, and our knowledge of lives lived before and after us, that gives humans the desire for purpose in their lives. For us, life is not simply focused on our survival; we desire to know objective truths about the world. This knowledge of the cause of birth and the inevitability of death gives us the long-term motivations that helped shape our cultural evolution. Our mastery of time also gives us our history and the perspective that comes with understanding long-term cultural and environmental change – we are able to understand our lives and cultural tools and practices in this rich context, so we have more meaningful collective cultural knowledge to draw on.

Our body's regular routines of waking, eating and sleeping that tie our biological cycles to Earth's rotations help mesh our mind's perception of time with the universal clock of the cosmos. To anchor our lives in the physical world around us, we needed to calibrate our culturally driven lives to the objective reality, and we started with time. We tried to investigate time rationally – it would set our species on an entirely new trajectory.

Siffre's experiment, which launched the field of chronobiology, revealed that we wake and rest to a cycle of 24 hours and 31 minutes.

Our body's timekeeping is automatic, set by constantly oscillating neurons inside the brain's hypothalamus, which are usually corrected by sunlight to keep us on a 24-hour cycle.

Our mind's perception of time, on the other hand, needs to be learned. Babies live so completely in the immediate present that it takes them a few months even to develop an understanding of object permanence – that when something disappears out of sight it still exists elsewhere, and could return. However, we do possess an innate perception of time intervals – babies recognize the difference between 20-second and 40-second intervals, for example. And they understand rhythms even before birth, which helps develop their language skills. But they live timelessly, unable to anchor their experiences to actual events, or to imagine their future or past. Although they have an ability to learn, small babies cannot make long-term memories. It is not until children reach three to four years old that they are able to time travel, to escape mentally into an event and imagine their emotions there. This helps humans develop the skills to regulate their emotions, and means they can anticipate or fear events that do not yet exist. Mental time travel also allows us to plan, which was transformative for our species.

We time travel using memory – it is what allows us to have cumulative culture and keep track of our large social group. By conjuring up the past, we can remember what worked in similar scenarios and repeat this, without having to innovate from scratch. Just as importantly, our memory allows us to time travel forward and imagine the future. To do this, our brain's prediction system relies on a sophisticated type of memory, called episodic memory, which may be unique to humans.[4] Unlike most types of memory that are timeless, such as the ability to learn a new skill or remember facts, such as the capital of France, episodic memory allows us to travel forward or back in time to visit any chosen event. It personalizes and contextualizes our memories, allowing us to learn from our experiences in a nuanced way. We can combine different strands of emotive and analytic information to make better decisions about future scenarios. This evolved cognitive ability gives us an important survival advantage, enabling us to rapidly adapt to multiple environmental changes and predict upcoming ones, such as seasonal events and the availability of our foods.

Episodic memory, like language, relies on the cognitive connections between different brain regions. Scans show a distinctive network of activation across the brain when an episodic memory is constructed and recalled.[5] Apes do not have this ability,[6] but it had evolved in our ancestors by at least 1.6 million years ago, because paleoanthropologists have discovered stone tools from that time that had been carried many miles from their place of manufacture. This means that the tool-makers anticipated using them at a future time. Other primates don't plan ahead: when they acquire a surplus of food, they discard what they don't want at that moment, even if their experience tells them they will go hungry later. They cannot simulate a world that's not the world they are in right now. Food-storing animals, such as squirrels, rely on behavioural instinct rather than conscious decision making.

Our experience of time is actively created by the mind, by our memories, emotions and the idea that time is out there somewhere connected to space. This 'mind time', our internal sensation of time, is central to our experience of reality. For most of us, time passes as a flow, like a river: behind us are the certain events that have occurred; in front of us are the nebulous uncertain events of the future. In the past few decades, a range of fascinating experiments have shown that emotions, fear, age, isolation, body temperature, rejection and attention can all affect our perception of how fast time flows.[7]

We need to calibrate our mind time with objective real time in order to make sense of the world and our place in it. Survival for our ancestors, including finding shelter, hunting, farming and travelling, was highly dependent on days and seasons. Our cultural calendar emerged from this: rituals, ceremonies and feasts evolved to mark particularly auspicious events and times of deep social vulnerability, such as when the days grew shortest over winter – and, in turn, timekeeping was essential for anchoring these cultural activities to nature's clock.

The most reliable clock our ancestors had was in the sky – our ancestors began to map the stars, finding meaning in their knowledge of its workings. Although the heavenly bodies rotate through the night, and swing back and forth with the passing seasons, their positions remain fixed relative to each other (as far as we can see from Earth), and they reliably pass through the same sequence of rise and set every year. People made small, portable lunar calendars on pieces of stone, bone, or antler so that they could be easily carried on extended journeys, such as hunting trips that lasted many weeks, and seasonal migrations. One such device, a small, carved eagle bone dating back at least 30,000 years, was found in a richly painted cave complex above the Tardoire river in the Dordogne region of France. It has marks and notches scratched into it, including circles, crescents and arcs representing the waxing and waning of the moon's phases over a 14-day period. An even older version, this time a 38,000-year-old tiny mammoth-ivory tablet, was found in a cave in Germany's Ach Valley. It has an etched catlike human figure with his arms and legs outstretched and a sword between his legs, which experts have interpreted as the Orion constellation, and 86 clear notches marked on the sides and back of the tablet, which may have a fertility role.

The ambitious astronomical maps found at the Lascaux cave complex in France are remarkable, including a lunar map, dating back 17,000 years, that depicts the 29-day cycle of Earth's satellite in groups of dots and squares. Above these dots is a row of 13 dots, representing the quarter moons – counting from the first winter rising of the Pleiades constellation to 13 brings the time when the horses are pregnant and easy to hunt. Star maps of the constellations appear throughout the caves among wonderful paintings of other important phenomena.[8] The people who made these detailed cosmic maps were scientists; they were making sense of their world through objective measurements of natural phenomena. Lascaux may well have been a prehistoric planetarium for charting the stars.

Archaeologists looking anew at prehistoric cave paintings are discovering star maps across Europe, where people were making mathematical and scientific observations of the cosmos. Our hunter-gatherer ancestors developed a range of techniques to navigate space and time, charting the night skies and marking the sun's journey in fluctuating shadow-lengths to create astronomical clocks of increasing intricacy. Stonehenge may also have been an observatory to track the motion of the sun, moon and stars. Certainly its builders were knowledgeable astronomers, mathematicians and architects – how else to explain the precisely calculated positioning of several enormous stones so that the main axis is exactly in alignment with sunrise at the midsummer solstice? Across the Irish Sea, at Newgrange in the Boyne Valley, is an even older tomb monument of astronomical mastery, involving some 2,000 quartz slabs quarried and transported from 80 kilometres away. For most of the time, the deep burial chamber and its 20-metre-long access passageway are pitch black, but at sunrise on the winter solstice, a shaft of sunlight pierces a small opening above the main entrance, known as the 'roof box', illuminating the back of the chamber. The designers of this important monument understood perfectly the angles, the position and movements of the sun through time.

Considerable time and energy was needed just to construct these monumental edifices – they were collaborative community efforts. But more than that, they also required keen astronomical observation, learned knowledge and accurate prediction, which take generations to acquire. The people who built them – we find similar constructions from Kenya[9] to Australia[10] – were investing in scientific infrastructure, judging the knowledge they hoped to gain worth it.

Astronomical knowledge was a cultural-environmental adaptation to help our ancestors survive seasonality and predict the availability of food sources, and observations are embedded in stories and songs, such as the Aboriginal songlines that are transmitted across generations. Take the Wergaia traditions of western Victoria: A terrible drought left people starving, when a woman named Marpeankurric set out to seek food for the group. After much searching, she found an ant nest and dug up thousands of nutritious ant larvae, called bittur, which sustained her people through the winter. When she died, she became the star Arcturus. Now, when Marpeankurric rises, she is telling the community that it is time to harvest the ant larvae.

Other Aboriginal stories describe how eclipses work, how the planets move differently from the stars, and explain the relationship between the moon and tides.[11] Certain constellations, such as the Pleiades, became culturally important across the world. The Pleiades is useful because it's a distinctive reference point: the seven stars are close together and bright, and always rise over the horizon at the same time each year – this has also made the number 7 auspicious. In the Americas, the Maya and Inca associated the Pleiades with abundance because it returns each year at harvest time, and they built solar and nocturnal observatories to follow it. For the Zuni of New Mexico, the Pleiades are the Seed Stars because they appear at the start of sowing season. The north African Berbers use the Pleiades to mark the changeover of the hot and cold seasons, and in Ancient Greece they heralded the start of the safe sailing season in the Mediterranean.

Astronomy was an essential cultural tool for navigation, too. While many animals have genetically evolved mechanisms to navigate, whether by moonlight or magnetic fields, human navigation relies almost entirely on our culturally evolved ability to invent maps in our minds. These maps, which may be transmitted in stories, were based on landscape details and the positions of the astronomical players above. The Polynesians evolved an extraordinary mental 'star compass' to track the movements of around 220 stars. Memorizing a sequence of rising and setting stars, and keeping track of their speed, direction and time, enabled these expert seafarers to chart a star path through the night and command territory across the vast Pacific Ocean.

By aligning our mental clocks with natural cycles and events, and observing and predicting patterns, we were able to explore our planet's entirety and journey temporally, through our lives and the lives of others. Time gave our ancestors a reference grid, a language to map their lived position in space, and this language had practical applications: it allowed us to meet, make future exchanges, and discuss past and planned events. In this way, our use of time reduced the entropy – the chaos of chance – in our lives, and lowered the energy costs in our activities. For instance, storing food for times of shortage was particularly useful once we left the biologically rich tropics for wintry latitudes. Many animals genetically evolved to do this innately, but here, again, cultural evolution allowed us to adapt to seasonal variation in food faster.

Our concept of time evolved to help organize our societies. In using it, we relied not on subjective norms but on measurable objective norms that could be agreed across tribes. As our societies grew in size and complexity, more precise calendars were sought and the art of time-setting became an important specialism – astronomical experts were prestigious and highly sought after in all cultures, and the celebration of their skills at predicting certain events, such as harvests, was popularly extended to other areas. They were often regarded as magicians, able not just to predict the future but also to change it.

Meanwhile, the pressure grew to standardize time. For millennia, there has been little agreement on when a day starts and ends, how many months there are in a year and even how many hours in a day.[12] The problem with the celestial cycles is neither the number of days in the lunar cycle nor the number of lunar cycles in the year is a nice round number, or even a whole number. The lunar month is 29.5306 days long, the average solar year is 365.2422 days long; dividing one by the other gives a rather unsatisfactory 12.3683 months in a year. Different astronomers from around the world tried every conceivable way to reconcile these difficulties and produce a handy calendar that could be used by the masses, by religious clerics and by civil servants, and which would be reliably accurate each year.[13]

The Romans moved New Year's Day from March to January, which was then followed gradually by other calendars – Britain adopted 1 January only in 1752 – and, four centuries after Jesus died, the Christian empire reset the start of the calendar to his estimated birth date (1 BC was followed by 1 AD because the concept of zero hadn't been invented yet).[14] Time is relative but is seen as a quantifiable resource: when it was announced in England that 2 September 1752 would be followed by 14 September to update the calendar in line with most of Europe,[15] there were riots in London and Bristol over the 'lost' days. Now, the Gregorian calendar is used globally. However, the management of months into weeks also varied widely across societies – post-revolutionary France attempted to introduce a ten-day week as recently as 1792.[16]

The wide variety of ways we have parsed our societies' time, while experiencing and accurately measuring the same objective celestial time, is revealing. It shows that although scientific advances may add to our knowledge, the way we interpret and use that information is dependent on cultural norms and sociopolitical needs. The Julian calendar, made by mathematicians, astronomers and philosophers, and used by the Romans from 45 BCE, may have marked a transition in the European popular perception of time from cyclical to a linear progression. This was a profound change that started to disassociate the measurement of time from the celestial cycles, and paved the way for other abstract ideas, such as mathematics.

The Romans were the first culture to have their lives meted out by time in a way that we would recognize in the industrial West. Sundials had become sophisticated and were everywhere in public spaces and privately owned. By the first-century BCE, the Roman architect Vitruvius could list 13 different types of sundials – and this was nearly two centuries after the playwright Plautus had cursed: 'Confound him who set up a sundial in this place to cut and hack my days so wretchedly into small portions!'

However, sundials meant the length of an hour varied with the seasons, because daylight began at dawn and ended at dusk. The Romans inherited their 24-hour day from the Babylonians, whose 60-base number system[17] was handily divisible by 2, 3, 4, 5, 6 and 12, in contrast to our base-10 system. But 12 hours of day and 12 of night meant that a Roman hour ranged from 75 minutes in summer to 45 minutes in wintertime. Gravity clocks avoided this problem in some situations: water clocks[18] were used in Roman law courts to moderate the time that each lawyer could speak for[19] – something that is surely due a revival in court and political debates.

Technology evolved to keep society in synch with the objectively measurable rhythm of the cosmos, but whereas in our deep past timekeeping had a survival benefit – alerting us to food availability, for instance – now it was overwhelmingly driven by subjective social norms. Christian clerics were extremely invested in astronomy to establish the dates of the solstices and equinoxes on which the complicated Easter calculations depend. The politics of the Christian calendar reveal our complicated relationship with time, and how we build our cultural norms around our interpretations of it. Easter, the most important festival in the Christian year, only started being celebrated in the second century CE, and morphed out of the pagan celebrations of spring.[20] Christians believe Jesus was resurrected three days after he celebrated Jewish Passover (Good Friday), which Jews mark on the fifteenth day of Nisan (around April) in their calendar – this corresponds to around the first full moon of spring, but the Jewish calendar has a leap month rather than leap day, so Passover moves around from year to year. Christians wanted Easter to fall on their holy day, Sunday, and they also wanted to make sure their new religion was distinct from Judaism and ensure that their holiday would never coincide with Passover. This may seem strange, given the integral nature of the Passover feast to the Easter story, but such is religious politics. In the end, the decision was to date Easter to the first Sunday after the first full moon after the spring equinox, unless that moon fell on a Sunday, in which case, Easter would be delayed to the following Sunday. A complex system of astronomical and mathematical calculations to model the motion of the moon, sun and stars was required to date the vernal equinox into the future, so Christian clerics led and supported observational research in astronomy for centuries. And the Christian calendar remains lunisolar – keeping pace with the seasons, but celebrating certain holidays according to the phase of the moon.

For Islam, too, accurate timekeeping, calendars and almanacs were very important, because Muslims need to pray at five specific times a day and must know the direction of Mecca. This drove astronomy in the medieval Islamic empire, and one of the most important scientific instruments to be improved during this period was the astrolabe, a multipurpose device for measuring the position of celestial bodies using angles or gradients. It could be used to work out the time, for surveying land and to calculate latitude[21] for navigation at sea.

Timekeeping finally became divorced from the motions of the heavens in the fourteenth century with the invention of the escapement mechanism, a device to regulate the rotation of a wheel pulled by a falling weight. Such a wheel could operate a gear mechanism, striking a bell on the hour (the word 'clock' comes from the French for 'bell'). The tick-tock of the clock's escapement would become the voice of time into my lifetime. Mechanical clocks meant that the length of an hour, no longer measured against a sundial, remained constant with the seasons.[22]

In the English cathedral city of Wells, a beautiful fourteenth-century clock, its face decorated with the period's geocentric version of the cosmos, displays the day of the lunar month and the phase of the moon, and still marks out the quarter hours with an elaborate display. With the construction of monumental public clocks, time became a valuable commodity – people could hear its passing, and the more accurately they marked time, the more dominion they gave it over them. This new quantitative cultural evolution in our approach to time extended to other areas, too. There was new precision in weights and measures, monetary standards, double-entry bookkeeping, perspective in painting and polyphonic music. Our attitude to the world had changed in western Europe and we began to perceive and classify things numerically – a trend in social norms that has only become more pervasive, and perhaps obsessional. 'Wasting' time began to be seen as not just foolish but sinful.

Time mastery grew out of larger, complex societies, but it also made such societies possible, because it eased trade and lowered the energy costs of our interactions and activities by removing uncertainty ('wasted' time); as states became more complex, time increasingly dominated all aspects of life.

The new perception of time as something that could be ticked off in equal minutes, let alone seconds, revolutionized society, leading to a scheduled world. Clocks[23] became universal in town squares, workplaces and homes, and people began wearing pocket watches.[24] Business time was measured not in how long it took to complete a task, but in how many man or woman hours were worked. Previously, people woke at dawn, worked according to their own schedule of demands, and then rested at night. Now machines, industrial mills and looms set the pace and depended on people working set hours, starting and finishing at the same time. People clocked on and off. Time became money – not passed but spent. The regimented diktat of clocks dramatically altered human lifestyles. Knowing the time became essential, and what time it was no longer had much to do with nature's cycles. As the Romantics complained in their poetry and their art, humanity had stripped the day from its earthly origins and had reset it to the rhythm of the working day.

Our invention of time changed the human environment into one meted out by time, and that changed our culture and biology. It is likely that we were once more knowing of our natural rhythms – more in tune with the fluctuating cycles of our bodies – when more of our external cues and stimuli were produced by the natural world. Plautus's moaning, two millennia ago, is revealing – once time begins to own you, it changes how you live. Young children, as animals, adrift in a timeless world, can be lost in play for hours or minutes, regulated only by their biological cues of hunger or tiredness. As they learn the social norms of how their cultural time corresponds to real, objective time, their mind time changes: in some cultures, time remains fairly relaxed even for adults, whereas in industrialized societies not being busy can produce feelings of guilt. In English, in societies where time dominates people's lives and is finely calibrated, the word 'time' is used more often than any other noun. The Amondawa people of the Amazon, by contrast, have no word for time, month or year.

Since 1972, all humans have officially followed a universally agreed time, but cultural time remains diverse. In an experiment in the 1990s, social psychologist Robert Levine compared the pace of life in 31 countries around the world, looking at average walking speed, accuracy of clocks and efficiency (how long it took to purchase a stamp in a post office). The results[25] confirmed that the world operates on completely different tempos as well as time zones. Those that were most fast-paced also had the strongest economies; cities were faster than rural areas, and the tropics were slower than countries in higher latitudes. Levine noted that in the western highlands of New Guinea, the Kapauku people do not work two consecutive days, and the !Kung hunter-gatherers of southern Africa only work 2.5 days a week for around six hours a day. Many parts of the world are simply not in a hurry – buses don't run to a timetable but leave when they are full. In India, a man can choose to give up working and spend his time seeking spiritual enlightenment and mystical insight. This is socially acceptable and people bring food to support his 'journey'; in the West, he would risk arrest for vagrancy. Westerners are suspicious and disapproving of people who don't use their time productively, and workers put a lot of effort into looking busy. European culture is full of images of skulls, a reminder that time is precious, you only live once.

In the 1920s, radio engineers working at the Bell Labs discovered that quartz crystals expand and contract in equal time when an electric current is applied, and, unlike pendulum clocks, quartz devices were undisturbed by atmospheric changes in humidity and temperature, or movement. Cheap and accurate new watches flooded the market, increasing the accuracy of timekeeping by several orders of magnitude. What the new precision of quartz time revealed was that the days, as meted out by the sun, Earth and moon, were not as accurate timepieces as we had thought, for day lengths fluctuated according to the pull of tides, the motion of the molten core at Earth's heart and even wind patterns. This disparity became even starker in the 1960s, once we began using an atomic clock pulsating to the exact rhythm of electrons and time became accurate down to the scale of nanoseconds, a thousand times more accurate than the microseconds of quartz.

A day stopped being the time it took for Earth to complete a rotation, but instead became 86,400 atomic seconds. But the time invented by human culture, dependent on the physics of the universe, still obeys the time set by earthly biology: the world's atomic clocks are reset every year to marry with shifts in Earth's orbit, so that atomic time and solar rhythms don't wander too far out of sync. Every year, the time lords at the International Earth Rotation and Reference Systems Service decide whether to add a 'leap second', depending on how much time the Earth has lost that year due to fluctuations of its rotational speed. If leap seconds[26] weren't used, our atomic-based international time would be obviously out of whack with Earth time in a matter of decades.

Our move to disassociate cultural time from biological time has flooded our homes and cities with artificial lights, completing our divorce from the dawn and dusk cycles of the natural world, and confusing animals and plants into untimely dawn choruses or unseasonal flowering. The result is a mismatch between the clocks in our cells and our cultural interpretation of the atomic time beamed by satellites to our smartphones. We work late into the evenings, rise early in darkness, and those of us who are office-bound at high latitudes may not see the sun at all for weeks over wintertime. We live, many of us, in a perpetual jetlag, which impacts our health, including an increase in cancers and depression, and also our relationship with the natural world.

The geographical distances that once isolated populations for evolutionary-scale time have shrunk as travel times become shorter and the time it takes to communicate collapses to an instant. Yet living in this faster world, with an accurately ticking clock, has given us a new perspective on time. We now understand ourselves and our place in the universe in the context of a timeline that starts with the Big Bang. Each new insight, from the dating of Earth's formation, to our ancestral relatedness to all life on Earth, has updated our self-knowledge and shaken our social identity and cultural beliefs. In 1837, Charles Darwin sketched a branching tree of lines in his notebook under the words 'I think',[27] as he depicted his theory of how life had evolved over eons. Little over a century later, a pencil sketch by Francis Crick (based on Rosalind Franklin's X-ray diffraction image) identified the double-helix configuration of atoms that make up a molecule of DNA, the beautifully simple 'essence of life' that allows genetic information to be transferred between life forms. Just as the stripes of accumulated layers of sediment laid down in rock mark geological time, so our DNA is a genetic clock[28] of life itself.

H. G. Wells published *The Time Machine* in 1895, a decade before Einstein published his theory of special relativity;[29] for the first time the possibility that humans would have complete control over time was beginning to feel mathematically possible. The irony is that for all our new expertise with time we nonetheless find it as difficult as ever to plan for humanity's future, and to imagine a world in which we are dead. This is perhaps a cognitive failure of biology, and it is certainly a cultural failing.

13

Reason

On the Greek coast at the foot of Mount Parnassus is a sacred crack in the rock – it has hosted a shrine for at least 3,500 years, and the cave leads, according to Zeus, to the navel of the Earth herself. Various stories have been ascribed to this geological fissure at Delphi, including that it was the place where the god Apollo slayed the dragon, Python, whose rotting body exuded a sweet smell that drew people to the site.

It is said that a goat herder named Coretas noticed one day that one of his goats, which had stumbled into the cave, was behaving strangely. Growing curious, Coretas entered the chasm and found himself filled with a divine presence and discovered he could see into the past and also the future. He could fly freely on the wings of Time itself, escaping our limited human perspective.

The story spread and many people started visiting the site to experience convulsions and inspirational trances, and some were said to disappear into the cleft due to their frenzied state. A shrine to the goddess Gaia was erected there and the villagers chose a single young woman as liaison for the divine inspirations. In time, the shrine was rededicated to Apollo, and the Oracle then spoke on behalf of all the gods. Before a prophetic session, the timings of which were determined astronomically according to the constellations, the Oracle would descend deep into the cavern and breathe in the sweet sacred Python fumes. She would then enter a trance and begin to rave, producing ecstatic speeches that would be received with awe and reverence.

The Oracle's knowledge of the future and sublime predictive powers were irresistible to mortals, who were immutably stuck in the present. She had enormous influence for centuries, and emperors would consult her on matters from love to war. Her visions would change fortunes and dictate life-and-death decisions. The Oracle was all-knowing.

The Oracle personifies a profound evolutionary drive for prediction. The more accurately we can predict the future, the better our survival chances and that of our descendants. Since we cannot time travel we have evolved other tools to see into the future – we use reputational information to navigate our social world, for instance. But navigating our physical environment requires new ways of knowing. To make better predictions about our world, we need to better understand it by exploring how it works, by observing and measuring. Wonder motivates us to look outside of learned subjective knowledge and to interrogate the world rationally, to seek meaning in objective truths. Our curiosity leads us to experiment and innovate: it makes us scientists, explorers and engineers.

Science is built on making predictions and testing those theories, and the knowledge that derives then allows us to make more accurate and diverse predictions, and accelerates our technological advancement. This type of cultural evolution, which uses the human-evolved capability for critical thinking, rationality and reason to innovate solutions rather than iterative copying, is often in conflict with our subjective knowledge. Nonetheless, it has propelled our species' cultural complexity and dominates social norms for learning. Most people now believe they make better decisions about the future using reason. However, the oracles we choose may not always be the rational ones.

Knowledge is the substance – the unit – of cultural evolution. As knowledge is transmitted – copied – between people and over generations, little tweaks occur that may confer a survival advantage or be socially attractive, and these result in adaptive improvements over time: culture evolves, in other words. The process is analogous to the mutations that occur in genetic evolution. However, in human cultural evolution, intelligent design also plays an important role. Deliberate innovation by individuals can dramatically speed the pace of cultural change. Although the lone genius producing sudden and startling inventions is largely a myth,[1] we do innovate within a cultural cradle: our inventions are almost always based on insights made by others, often making novel connections between existing things. Such break-throughs, which are the result not of selection of copying errors but of original invention, enable cultural evolution to progress in leaps rather than by increments. Purposeful invention speeds the pace to complexity.

Innovation is widely used by animals and correlates with increasing brain size. There are countless examples of biologists observing new animal behaviours, including the English songbirds that figured out how to sup the cream from milk bottles by pecking through their foil caps, and roof-sledging crows.[2] Innovation helps animals adapt more quickly than via the slow pace of evolutionary change to innate behaviours. One study[3] found that innovative species of birds were significantly more likely to survive when moved by humans to new locations, for example. Our ancestors' ability to innovate would have been indispensable as they migrated relatively rapidly around the globe.

This way of knowing, through autonomous trial and error experimentation, is perhaps our most primitive. After all, the brain evolved as a prediction system to enhance its owner's survival – the more we interact with the world, the better our predictive capability. Babies and small children explore their environment through their senses, tasting and observing objects to discover that objects that collide can produce an acceleration, such as a foot and a ball, or that ice is colder than water. But the human encultured brain evolved to prioritize copycat social learning, which wins out over inventiveness, as we've seen, because it is much more efficient to copy successful outcomes and to make predictions based on a collective of others' experiences than be limited to our own narrow range. Innovation is a risky strategy with a high failure rate,[4] so it is used relatively rarely. One analysis[5] of how culture evolves in real-world situations used online programming contests and found that improvements were overwhelmingly made through iterative tweaks while copying the best-performing solutions, and only rarely through innovative leaps – the ratio of mutations to inventions was 16:1.

Although innovation is comparatively rare, it's important, because without it, our culture would tend to reduce in diversity over time, as populations focused on copying and refining only the high-performance solutions. That would leave societies without sufficient adaptive solutions and therefore vulnerable to crises, such as rapid environmental change. Together, our two cultural evolutionary processes generate a cascade of new possibilities and functionality in our collective intelligence. Deliberate modifications are included in our cumulative culture, where they are subject to the same selection pressures, and the best solutions are spread through the population through high-fidelity copying.

Innovations, though, just like advances made through copying, are built on the foundations of our collective knowledge, even if the step-changes are bigger. Once the wheel had been invented, it became easier to imagine the potter's wheel, wagons, war chariots, wheelbarrows, gears and waterwheels. Technological invention in particular, because it relies on the laws of physics and biology, accelerates with accumulated scientific knowledge. This means a rational understanding of the world based on experimentation and objective measurements. As this type of intellectual culture grew, so our innovations increased. Intelligent design in the process of cumulative cultural evolution thus works like a ratchet[6] – a certain level of cultural complexity has to be reached before the innovation is possible, but once the insight is made, society is ratcheted up in an acceleration of progress.

Mathematics, for example, took off with the invention of zero. The first evidence of written mathematics comes from the ancient Sumerians of Mesopotamia, 5,000 years ago, who developed the science of numbers and measurements, multiplication tables and geometry, which the Babylonians and Greeks expanded upon, making incremental progress. Once the number zero had been invented, in the seventh century CE, it was suddenly (with Arabic numerals) easy to distinguish between 1,000 and 10,000, using zero as the decimal placeholder. Higher mathematics became possible, as well as lots of practical applications, including simple financial accounting. Zero also allowed the infinite precision of digital fractions (after a decimal point), which enabled thinkers like Newton to develop new laws of physics. (Dogmatic Christians, who argued that, since God was in everything, zero was satanic, attempted for 1,000 years to banish zero from Europe, but were unsuccessful.)

*

Over the past dozen centuries, as astronomers, philosophers, mathematicians and engineers explored different ways of knowing and
improved their understanding, they became better at predicting
events in the unknown realm of the future. Such predictions could be
tested – they didn't rely on an assertion by authority or belief in a
shared story, but were based on measurements and calculations using
rules that were objective and had true external values.

This is different from other areas of cultural evolution. There is no
logical hierarchy to subjective knowledge about the world: if you say
that a bride should wear white, and I say that white is for mourning
and brides should wear red, then it is purely a matter of opinion over
which you choose to believe. Our folklore differs between cultures.
There is no Eastern version of gravity that differs from the Western
version, though; there is no Western science – there is simply science.
This doesn't mean that the symbolic meaning we have sought in
events is lost. There remain many questions that science cannot
answer and for which we continue to seek explanations through
stories and other cultural interpretations, such as what is the meaning
of life, and what is consciousness? For some people, science is not the
right tool for these questions; others think that science will one day
answer these questions using reason; while a majority of people who
use and accept scientific arguments nevertheless believe in spiritual
explanations for other things and reconcile both ways of knowing
with ease.

The first general scientist[7] that we know of whose predictions could
compete with supernatural divinations, like the Oracle's, was Thales,[8]
who lived around 2,600 years ago in ancient Greece. He studied in
Egypt and Babylon, and on his return, he transformed the field of pure
maths, establishing, for example, that mathematical theorems must be
proven before they can be accepted as true. He also proposed rational
explanations for natural phenomena, such as the flooding of the Nile
and earthquakes, that debunked stories about angry gods. But it was
his evidence-based agricultural predictions that made him wealthy: by
studying weather patterns in the Miletus region of Ionia, he was able to
accurately forecast the following season's harvest. One winter, he calculated a bumper olive crop and placed small deposits to hire all of the
olive presses in Miletus for the next harvest. In summer, when the olive

growers realized a huge crop of olives was coming, they discovered that Thales had already taken all the presses, and he made a fortune by selling on his hiring rights to the growers.

The spread of knowledge and innovation through society is heavily influenced by social norms that stifle or encourage investigation. For the ancient Greeks, the idea of quests and questioning, of philosophizing and gaining knowledge through debate and observation were integral to the culture of intellectual life. Religious stories were enlivened and informed by new thinking and discovery; they decorated and enriched a culture of tolerant, thoughtful reason and rationality. But philosophical and scientific inquiry would become casualties of Christian dogma.

The fall of reason began with St Paul, in the years soon after Jesus's death.[9] Paul was a zealous Jew who persecuted Christians before his own conversion. His Christian doctrine instructed that Greek philosophers were blinded by their own questioning approach and would go straight to hell. By the fourth century, under Emperor Theodosius, the Bible had become the last word on everything and it had become heretical to question anything. Rome went from being a relatively open, tolerant and pluralistic civilization toward a rule of fixed authority, whether that be the Bible, or the writings of Galen and Hippocrates in medicine, or Ptolemy in astronomy. This was a marked transition in the West from the ancient pagan world of philosophical reason to a static creed that not only unequivocally rejected scientific and rational thought but embraced a dogmatism that frequently and brutally punished those who did not satisfactorily conform.

The remarkable Greek mathematician, astronomer and philosopher Hypatia, one of the last great thinkers of ancient Alexandria, and a rare woman academic of the time, was one such casualty. Hypatia ran popular lectures on Plato and Aristotle and taught students in mathematics and astronomy, including instructing them on how to design an astrolabe. She was a widely admired intellectual, and although she was in a relationship with a prominent Christian, she was not herself a convert. In 415, a mob of Christian zealots led by Peter the Lector accosted her carriage as it drove through the city streets, dragged Hypatia from it, and took her into a church. There, they stripped her and beat her to death with roofing tiles. They then tore her body apart and burned it.[10] She was killed for having an inquiring mind.

Higher levels of religious intolerance are generally associated with reduced creativity and technological innovation.[11] As norms changed to favour faithful copying over innovative thought, the collective culture shrank. During more tolerant Classical times, learning was valued and the entire wealthy upper class and trading middle class were literate, and part of a vast, active network. From the fifth century, the Western Roman Empire had fallen, literacy levels plummeted outside of the Church, and early medieval Europe became a feudal society, a period referred to as the Dark Ages for its lack of scientific and technological innovation – the clerics relied on manuscripts from antiquity and were not in the business of scientific research. The networks had broken, and in many ways, cultural evolution regressed in complexity. During such dark ages, ways of knowing were limited through population crashes, isolation, and social norms and institutions that restricted the flow of information, intentionally slowing cultural evolution.

Further east, though, norms were very different. Literacy was highly valued, for men and women, and Muslim scholars adopted Greek and Roman teachings in science and medicine and added Persian and Indian traditions. By the eighth century, Baghdad had become a global centre of learning under the Abbasid dynasty, whose empire stretched from Spain to China and from Yemen to the western Sahara. Baghdad,[12] a city of 2 million people at a crossroads between Europe, Africa and Asia, was able to draw on a wide range of cultures, ideas and experiences. This connectivity, open-minded tolerance of different ideas, and the emphasis on learning, made the so-called Golden Age of Islam a scientific bastion. For over 700 years, the international language of science was Arabic.

Innovation often arises from people recognizing useful connections between different ideas, and for that, standardization helps – the use of a common Arabic language meant that knowledge could be spread more widely so those 'aha' moments were more likely to occur. It was in Baghdad that Arabs learned from Chinese prisoners of war how to make paper, generating a cheaper, faster way of disseminating information than papyrus or parchment, which were used elsewhere. With paper and a new and easier writing system, information was democratized to the extent that it became possible to make a living from simply writing and selling books.[13] Under the patronage of Arabic-Persian Caliph Abū Ja'far al-Ma'mūn, an ambitious project to harvest all the world's knowledge was begun – scholars from other cultures were welcomed and Arabic emissaries were sent with funds to the far edges of the known world to bring back documents and manuscripts. Defeated foreign rulers would be required to settle the terms of surrender to him with books from their libraries rather than in gold, such was the value of knowledge and information. In a grand translation project, these works were then written into Arabic to be kept and studied in the House of Wisdom (Bayt al-Hikma), creating a library that aimed to rival the destroyed one at Alexandria 1,000 years earlier. This initiative alone preserved swathes of ancient knowledge that would otherwise have been lost to history – just as global dispersal can help species to survive an extinction, enabling small populations to restore genetic diversity, so libraries, monasteries, and practising communities can protect against cultural extinction.

It took a thousand years for Europe to free itself from the effects of Theodosius's decision. The trade in ideas from the Islamic world to the Christian kick-started the European Renaissance in science and exploration, as new populations rediscovered ancient thinkers.[14] (Ironically, this also resulted in pushing the Islamic world into a cultural regression from which it is still recovering.) The Church led and still controlled[15] much of early scientific search for knowledge in the West, but this changed in the mid-fifteenth century with Johannes Gutenberg's invention of moveable type, which, in combination with Europe's new use of paper, led to the popular printing press. Printing disseminated information in standardized texts, so all over Europe people were reading the same thing, and could make easier comparisons and cross-references.[16] Information now had a much greater reach, especially when Aldus Manutius, a Venetian printer and publisher, invented the smaller, cheaper octavo volume.[17] The democratizing of the printed word[18] inaugurated a set of sweeping changes to social norms, encouraging exploration, experimentation and inquiry.

'Dare to know' became a motto for natural philosophers, and wonder went from being a sign of ignorance to an applauded desire for knowledge. Late-fifteenth-century scholars[19] questioned whether something was true just because it was written in an old book, and started to argue that the most reliable way to get knowledge was from direct experience: to look for yourself. In the 1660s the word 'fact' entered common use.[20]

The more someone uses rational thought processes, the more practised they become. Through social interaction in the course of our development, we not only acquire facts about the world and how to deal with them, we also build the cognitive processes that make 'fact inheritance' possible.[21] In other words, cultural learning is itself culturally inherited. Social norms that evolve to create institutions that generate a larger collective brain also make us individually smarter[22] – it's not how intelligent we are born but how social we are that decides how clever our culture's inventions are.[23] That's one reason why universities, which concentrate large numbers of people from different cultural developing baths, foster ideas and technological innovation.

Scientific logical reasoning is a cognitive processing tool, a way of looking at and understanding the world. People who undergo their social learning in the cultural developing bath of rational thinking generate the cognitive tools that lead them to seek knowledge and explanations scientifically. These are biological changes to their brains[24] as a result of a cultural developing bath in which rational ideas are given authority. Such people are more likely to question the status quo, and so their cultures accelerate technological and scientific change,[25] as well as social change.[26]

Our cognitive tools, such as literacy and the idea of time, also contribute to technological advancement. In societies where children are taught literacy, they are better able to develop arguments that build upon others' and progress. Different ways of thinking, when practised, biologically shape the brain's cognition pathways, and this sometimes involves a cognitive compromise. Higher mathematics involves manipulating numbers and symbols in more complex ways and cumulatively over many equations, so literate mathematics – writing down operations – developed to enable that. But it is a less efficient way of calculating sums than bead counting on an abacus, which has been used for nearly 5,000 years to add and subtract numbers. Trained abacus users can solve multiple math sums faster than someone operating a digital calculator. In parts of the world where the abacus is regularly used, adults doing calculations in their heads mentally construct a virtual abacus to slide beads that also solves mental arithmetic problems faster than the same challenge given to Western undergraduates, who rely on their linguistic cognition (number words) to make the calculations.[27]

The physicality of scientific endeavour is often forgotten, but we are never minds floating in an intellectual domain divorced from our baser, corporeal selves. Our bodies have always informed our minds and they have evolved in conjunction – we understood the world first through our senses and by grappling with the stuff of nature. Our cultural evolution to rationality and science was famously led by elite scientists, such as Isaac Newton and Charles Darwin, but it relied on a grassroots obsession with data, instruments and measurements. It is as much the manipulators of stuff, the artisans, mechanics and engineers who produced the scientific discoveries of the past 500 years, as the philosophers and thinkers; indeed, one reason that Europe led this enlightenment was the pragmatic practicality of many of the thinkers, unlike the purely intellectual scientific culture of, say, China. Many of the vital players driving British science and the industrial revolution didn't go to Oxford or Cambridge but had artisanal training; for instance, John Harrison, who solved the problem of determining longitude at sea, was a self-taught carpenter and clockmaker, while James Watt, who invented the improved steam engine,[28] was an instrument maker.

Developments in science, technology, financial systems and others all drive each other and accelerate the quest for new knowledge, but all rely on states, institutions and social norms. Science is effectively a long-term public-goods project, and depends on support from patrons, including religious institutions that want better solstice data, businesses that need improved harvest prediction and government bodies that need to calculate taxes. While it takes some time for scientific concepts to be accepted by scientists in the field and then by the population generally, the tools and technologies that arise as part of scientific endeavour are taken up faster, and can be transformative to other fields. Science is collaborative, and standardization in time and other measurements was a vital consensus-building process that created stability in technological systems and enabled global trade and exchange in ideas and parts. Standardization thus acts as a cultural lever accelerating technological evolution.

Science works by disproving theories. The complicated geometric acrobatics that Ptolemy needed to make his Earth-centric model of the universe plausible was never very satisfactory; it was just the best model scientists had – until it was successfully challenged by a better theory. However, new objective inquiries and discoveries about how the world works don't suddenly shine a light into the gloom of collective ignorance. Scientific theories can be hard to distinguish from subjective explanations, and people learn their beliefs by copying others. We are not, most of us, living in such denial that we can truly hold belief in opposing scientific and religious stories. When our understanding was that Earth was the centre of the solar system, this model formed part of our religious and cultural narrative. When Earth was relegated to just another satellite of the sun, this was a paradigm shift for science and also radically shook our identity – the story of our special planet made by a god at the centre of the universe had to change. In time, most people shifted their perspective to the latest findings, and the religious narrative morphed or believers circumvented its contradictory explanation.

It can take centuries to change our understanding, especially if it is counter to our experience. From my perspective, the Earth is still and the sun appears to the east and revolves around the Earth in a daily arc, setting in the west. I had to learn that this was not true, and although intellectually I believe the heliocentric model, emotionally, I struggle: it feels as though the sun is moving through my day. As scientific explanations grow more complex still, that emotional disconnection becomes greater: I understand the basics of quantum mechanics, of gravity and magnetism from a mathematical perspective, but intuitively not at all. These are concepts that rule my life, and yet the understanding and acceptance I have for this knowledge is very different from other forms of cultural knowledge. Perhaps more alarmingly, studies show that most people routinely fail to understand the relationships between very big numbers, assuming, for example, that millions, billions and trillions[29] are equally spaced on a number line, which could have an important impact on how they view government policies and decisions.

Part of this is because we do not regularly manipulate things at this scale, so we don't have the same intuitive understanding of vast numbers that we do for numbers below 20.

Our embodiment informs our cognition. Our brains have evolved to construct for us a version of reality based on our models of perception, but this interpretation depends on a mixture of biology, cultural experiences, and environmental references. The way that two people perceive reality can be completely different, and our brain has to balance our experiential knowledge against our objective knowledge and make a judgement.

Optical illusions reveal how easily sensory information from our eyes can be misinterpreted by our brains to form an altered version of reality. Even when we know that optical trickery is at work, we can be fiercely defensive of our own perception of reality. In 2015, a journalist at Buzzfeed found a picture of a striped dress and posted it on the Internet with the caption: 'Guys please help me – is this dress white and gold, or blue and black? Me and my friends can't agree and we're freaking the f—k out.' Within hours, many thousands of other people were also freaking out over The Dress, and social networks were abuzz with angry tweets and outraged posts by people who couldn't countenance that others have a different perception of the world. There is nothing more sacred to us than our reality – our relationship between our internal thoughts, the outside world and our own bodies is fundamental to our sense of person; our grasp of reality is the handrail that keeps us from insanity.

In the past, when there were no scientific explanations for visions and other strange phenomena, such experiences were convincing evidence for gods. Recently, a team of scientists investigating the underlying geology at Delphi discovered two hidden fault lines that cross exactly under the ruined temple. Psychoactive gases seeping up through these tectonic cracks, including sweet-smelling ethylene, which produces feelings of aloof euphoria at light doses (and anaesthesia at higher doses), were probably responsible for the Oracle's visions. As we learn more about the brain's neuroreceptors and how we process visual data to create our model of reality, we are closer to understanding how different chemicals can lead to altered perceptions of reality and religious experiences.

Most of the way we build our perception of the world is done without thinking, subconsciously, by our brain constructing a version of reality from the limited amounts of information it receives. The neurologist Antonio Damasio describes a 'somatic processing' system in decision making, whereby the brain's ventromedial prefrontal cortex creates bodily signals, such as blood pressure changes or increased heart rate, to mark the unconscious decisions it makes (on the basis of past experience), and then the brain interprets these as gut decisions made before conscious reasoning has time to catch up. In the case of The Dress, there is some evidence that people who spend much of their time outside under natural light conditions are more likely to perceive it as white and gold, whereas those who spend longer inside are more likely to see it as blue and black. Babies, up until about four months old, do not suffer from this so-called perceptual constancy, and will see the correct colours. However, the human brain learns to override objective differences for subjective similitudes.[30]

Our cultural developing bath influences our biology to produce a version of reality – and this influences our political choices, our beliefs and our behaviours. Our social groups then reinforce our belief systems, through the tribal effects of discouraging or condemning out-group beliefs or by bolstering supportive opinions. Social media[31] is a great example of this so-called bubble effect. So do we believe our experiential version of reality or the objective version? After all, things that seem unquestionably right by one group, can appear crazy or wicked to another: take views on gun ownership, abortion rights, or gay marriage, among other topics. A few decades ago, it might have been working mothers, national service, or phrenology. We navigate the world by our common-sense perception, but that perception has blinded us to reality again and again. We have mistaken our sensorial intuitions for facts and have mistrusted processes and phenomena beyond the boundaries of what we can touch and feel with our limited senses – from evolution, which unfolds on scales of time too vast to be visible within a human lifetime, to quantum mechanics, which operates on subatomic scales imperceptible and almost inconceivable to the human observer.

Aristotle described humans as the 'rational animal', but all too often we behave irrationally. It was recently revealed that water utilities, which are equipped with the latest scientific technologies, including satellite imagery and geology expertise, also rely on water divining methods, such as dowsing, to detect leaks. When this became public knowledge, scientists were outraged that medieval bunkum was being used in the twenty-first century – as a customer, it's certainly galling to be paying for it – but it reveals that in our portfolio of culturally learned solutions, rational options are not always clearly delineated from the stories.

Critical thinking was our culturally invented tool to help us rationalize situations, enabling us to form sound beliefs and judgements. The problem is, the rational explanation is not always obvious and can require complex calculations or statistical analyses to determine. So, when we need to make fast decisions based on complex issues, we often go with our gut instinct. This is partly because it is cognitively less demanding and therefore more energy efficient. The Nobel Prize-winning economist Daniel Kahneman describes the two systems as 'thinking fast' (unconsciously, intuitively, effortlessly) and 'thinking slow' (consciously, analytically, effortfully); and explained that even if we think our decisions are mostly rational, most of the time we are thinking fast.

From an evolutionary perspective, thinking fast and emotionally makes sense – many individual survival situations require a speedy decision (if you stop to think about whether you can outrun a lion, you have already lost), and intuitive decisions are often unconsciously based on pattern recognition, environmental cues, or other biases that have a useful basis. Group survival is also often dependent on thinking fast – if a firefighter or warrior stops to think about her own safety before rushing to the aid of her fellows, she might decide the risk was too great, whereas if she is successful, group survival is improved. Athletes and other skilled performers, once they have learned and practised their technique, would be hampered if they had to consciously judge and think about their every move. Emotions are useful: after all, fear compels us to respond quickly to risks; anger amplifies our interactions to make threats more credible; guilt prevents us acting outside social norms and risking group cohesion; and so on. In one study,[32] pairs of students were asked to negotiate the allocation of sums of money between them. Some of these students were manipulated to enter the lab slightly angrier by being made to listen to irritating music. The angrier ones eventually left with more money.

Despite our culturally evolved norms for rationality and evidence-based decision making, our biological evolution has not caught up and our cognition continues to be emotionally led. The problem is not necessarily that we use the emotive part of our brain more than the rational in decision making, but that we are self-delusional. Even experts are prone to biases and these mean costly mistakes are made, and irrational prejudices are systemic in organizations where people believe themselves to be non-racist, non-sexist, and to hold the positions they do through skill rather than luck.

Our decision making is influenced by our biology and our social environment. Take the psychological and physiological influence of fear: it's been shown that people who vote more conservatively tend to have a bigger amygdala,[33] the brain's fear centre. In one study,[34] the more fear a three- or four-year-old showed during a lab study, the more conservative their political attitude was found to be 20 years later. The impact of fear is instant: when people with liberal attitudes experienced physical threat, during a study,[35] their political and social attitudes became more conservative, temporarily. Conservative politicians and electioneering exploit this, aiming to raise voters' fears of immigration by comparing immigrants to germs, for example, which targets our deep, biologically evolved motivations to avoid contamination and disease. In one study[36] during an H1N1 flu epidemic, researchers reminded people of the dangers of the flu virus and then asked them their attitudes towards immigration, after which they were asked whether they had been vaccinated against flu yet. Those who hadn't received their anti-flu shot were more likely to be anti-immigration than the ones who felt less threatened. But in a follow-up study, the researchers offered people a squirt of hand-sanitizer straight after the flu warning, and the immigration bias went away. Making people feel safe changes their voting decision to more liberal.[37] When researchers asked people to imagine themselves completely invulnerable to any harm, Republican voters became significantly more liberal in social attitudes to issues like abortion and immigration. Reason is suffused with emotion. And this also has implications for cultural complexity because, as several studies – looking at everything from artistic output to patents filed – show, the more conservative a society is and the more restrictive its norms, the less creative it is and the fewer innovations it produces.[38] Technology advances fastest in more liberal societies.[39]

Sometimes, going with your gut, rather than rationalizing, can actually produce the better outcome, because by tuning out noise in the prediction system, our irrational cognitive biases often work well for complex decisions that have an emotional component. For example, statistical models can be error-prone because they contain inherent biases, are incomplete, or are based on mathematically perfect scenarios that are incompatible with the messy real world. That's why many financial models failed to predict the 2008 financial crisis. The social implications of most decisions are also important factors in decision making. Banking insiders who were anxious about an impending crash stayed quiet to avoid the social cost of voicing a disparaged rational opinion. In very partisan situations, people who disobey the social norm by voting against the group majority risk ostracism. In such cases, therefore, it may be more rational for the individual to go against the evidence because we are motivated more by social cohesion and maintaining support networks than being objectively right.[40]

Tribal culture still affects how people see the world more than facts do. Take human-caused climate change, for which there is near-unanimous global scientific consensus. This divides Americans, but in an unlikely way: the more education that Democrats and Republicans have, the more their beliefs in climate change diverge. About 25 per cent of Republicans with only a high school education report being very worried about climate change. But among college-educated Republicans, that figure was just 8 per cent.[41] This may seem counter-intuitive, because better-educated Republicans are more likely to be aware of the scientific consensus. But in the realm of public opinion, climate change isn't a scientific issue, it's a political one. Climate change science is relatively new and technically complicated, and many Americans adopt the opinions of their tribal leaders, the political elites. Republican political elites are not science-minded. Even though better-educated Republicans may have more exposure to information about the science around climate change, they also have more exposure to partisan messages about it, and research shows this matters more.

As Jonathan Swift pointed out in 1720, 'Reasoning will never make a man correct an ill opinion, which by reasoning he never acquired.' Since we have culturally evolved to acquire our knowledge and beliefs primarily through high-fidelity copying of others rather than by invention (by looking at the evidence and deciding for ourselves), we are vulnerable to this problem of copying unreliable models. Worse still, because we have culturally learned to value rational explanations over subjective ones for scientific issues, we can be manipulated into believing the opinions we copy are rational, so it is harder to change them.

Often, the main role of reasoning in decision making is actually not to arrive at the decision but to be able to present the decision as something that's rational. Some psychologists believe we only use reason to retrospectively justify our decisions, and largely rely on unquestioned instincts to make choices. It may be that our unconscious instincts – despite our cognitive biases and prejudices – are more capable of rationality than our logical thought-processing minds. Few of us are able to fully separate our subjective and objective reasoning during decision making – this is one of the promises of artificial intelligence. AI is logical, but it is only as objective as the designer of its algorithms allows it to be. However, there are many decisions that are subjective for a reason. Evidence-based science can provide us with tools to make decisions based on measurable outcomes, but social norms – society's values – determine how we act. The evidence for the link between gun ownership and gun crime is undisputed statistically, and yet, in the United States, a powerful minority throws its hands up after every mass shooting and wonders what on Earth can be done about it.

Biologists believe we are the only primates to have evolved the concept of information that's untrue or different from what they know. That means that other primates can't conceive of states of the world that are decoupled from their current reality, and can't imagine other individuals thinking about the world in a different way. We, however, know that there are things we don't know, and that other people hold different opinions. From this, we often conclude that we are the rational ones, whereas those we disagree with are irrational. It would be safer to assume that they think just as rationally as we do, but have different goals, background beliefs and priorities.

It used to be accepted that science informs us of the objective reality of the world, whereas our subjective explanations inform us how to feel about it. But increasingly, science is beginning to explain our subjective responses too, the way our emotions arise and how they are manipulated, how our memories develop and can be fabricated. As we learn more about the workings of our human mind, and develop increasingly human-like artificial intelligence, will we, by demystifying the storytelling part of our consciousness, eventually reach a point where we can choose pure rationality? Perhaps.

The most powerful supercomputer we've invented, Summit,[42] is capable of performing 20 quadrillion calculations – which would take the human brain 63 billion years – in one second. We use it to predict that most pedestrian of preoccupations: our weather.

14

Homni

Texas, 12019: You begin your pilgrimage at dawn, trekking deep into the mountains. Once you arrive at its hidden entrance in an opening in the rock face, you find a jade door rimmed in stainless steel, and then a second steel door beyond it. These act as a kind of crude airlock, keeping out dust and wild animals. You rotate its round handles to let yourself in, and then seal the doors behind you. It is totally black. You head into the darkness of a tunnel a few hundred feet long. At the end there's the mildest hint of light on the floor. You look up. There is a tiny dot of light far away, at the top of a 500-foot-long vertical tunnel about 12 feet in diameter. You start climbing a continuous spiral staircase, winding up the outer rim of the tunnel, rising towards the very faint light overhead. The journey to the Clock in the mountain ends at the summit in light. It is the sun that powers its ringing below. You are the first of your kind to hear it, for the clock has not rung out since it was built, 10,000 years ago.[1]

Time is relative. Continents move at around the same speed as fingernails grow. We measure time according to how fast things occur and human cumulative cultural evolution is speeding up time. Eons are shrinking to a human lifetime as geological time, once calibrated in tens of millions of years, is now passing in a matter of decades. Cities, located days away from each other, can now be reached in hours; people can be contacted in a second. The rate of extinction is one thousand times the 'natural' rate; the human population redoubles.

We have travelled far in a heartbeat of planetary time. Fifty thousand years ago, 100 billion lives ago, we were just one of a handful of similar human species who had evolved on this Earth; now we are alone. It has

taken time for cultural complexity to ratchet up and deliver the technologies and social institutions we have today – time for our populations
to build – and for most of it, we were constrained by the hostile environment of the Pleistocene. It's been shown that in times of hardship and
food scarcity,[2] societies become more culturally conservative and the
frequency of innovations falls;[3] individuals also perform worse in tests
of logical and creative thinking, and favour emotional, rather than
rational, decision making.[4] Humans have spent 95 per cent of their
existence in such conditions and yet, as we have seen, our ancestors
achieved remarkable cultural complexity, even during the height of the
last ice age. In times of plenty, the same study found, cognitive performance leaps up for those tests, and 11,000 years ago, the environment
became far more favorable. Earth entered the stable, mild climate of the
Holocene epoch. It arrived too late for any other human species to
enjoy, but we flourished. The Holocene delivered the increase in available resources that enabled our ancestors' populations and trade
networks to grow, accelerating our cultural diversity and complexity.[5]

Every efficiency in how we operate in and with our environment – in
other words, any adaptation that improves the flow of energy –
improves our survival and accelerates our cultural evolution. Social
complexity, for instance, is limited by the amount of energy societies
can harness, so when societies relied on just human and animal labour,
the activities of states were usually limited to war, food and security,
although there are notable exceptions – Rome prospered for 900 years
fuelled mainly by slave energy.[6] With new energy sources, such as water
wheels, states expanded trade, and there was more power to be gained
from wealth than war. With energy from coal, a whole new complexity
of government evolved with increasing bureaucracies. Complex systems are synergic – in this case, from a complex energy distribution
system has emerged our modern industrialized society.

This is because the availability of energy is proportional to its
cheapness. Innovations cannot spread to the extent that they become
a complex system unless they are affordable, and for most of our history, energy has been expensive. Take lighting:[7] in 1800, the average
person used 1,100 lumen-hours per year; two centuries later, that was
13 million lumen-hours – average consumption of light had become
11,800 times greater. This is down to cost. In 1800, it took a person

60 hours of hard labour to produce enough weak tallow candlelight (from mutton fat) to use for two hours and 26 minutes every day for a year. The same amount of human labour would produce just 54 minutes of light from an incandescent bulb. But look at the cost difference: 1 million lumen-hours of artificial light cost £2.67 in Britain in 2006; in the fourteenth century, it would have cost the equivalent of £35,000.[8]

Energy has become cheaper through economies of scale, technological innovation and other efficiencies, and that has accelerated the economy. For tens of thousands of years during the Pleistocene, the global economy doubled every 250,000 years; in the Holocene, with agriculture, it doubled every 900 years; since 1950, it's doubled every 15 years. In line with this, human population has soared from 1 billion to 7.7 billion in the past 150 years. And where is this new vast population of humans living? In the concentrated, efficient social systems of cities, which cover just 3 per cent of the planet's land surface, but will soon house 75 per cent of the world's population.[9] Urbanization is increasing humanity's network density as never before, creating its own emergent properties, including new mélanges of genes and culture,[10] orchestrated health-care systems, and the first slowing of population expansion as people voluntarily limit their family size despite – or indeed because of – access to greater resources. A baby born today in London is more likely to survive to adulthood – perhaps living for a century – than at any time.[11] She can learn from the biggest and most interconnected human population and has access to the greatest cognitive and technological toolbox; as well as literacy, she will be familiar with wheels, springs, levers, fractions, evolution, money, democracy, infection control, perspective, and so on. This toolbox means that today's humans can solve a problem more efficiently than at any time in our history.[12] In recent decades, the so-called Great Acceleration[13] in human activity has delivered a rapid increase in population, globalization and technological innovation.

In this book I've shown how we are continually making ourselves through a triad of genetic, environmental and cultural evolution, and how we've become an extraordinary species capable of directing our own destiny. We are now, all of us, on the cusp of something quite exceptional. Humanity is becoming a superorganism. Let's call him *Homo omnis,* or Homni for short.

To understand Homni, let me first take you down into the soil to meet one of the most simple and ancient single-celled organisms, an amoeba called a slime mould. It evolved some 600 million years ago, and occupies soils across the world from Antarctica to the Arctic. For most of its life cycle, the cell lives the unexceptional life of most amoebae. But sometimes, these single cells gather in their thousands to create an organism, encased in its own slime, that can creep, crawl, pulsate, grow tentacles, and even negotiate a maze. Scientists describe these slime moulds as 'societies' because of the way that the individual amoebae work together toward a common cause, sometimes sacrificing themselves. If food is scarce in their soil patch, the amoebae will coalesce into a tendril that creeps up to the light. There, a portion of them will form a stalk above ground, by turning their bodies into hard cellulose in a process that kills them. The rest of the mould then climbs up the stalk and waits in a blob at the top for a passing animal onto which they can stick and thereby be transported to new soils.

The human brain is a bit like the mould, albeit not as independent or physically mobile. Each single brain cell, or neuron, cannot be described as sentient and yet, when all 100 billion neurons are networked together, the human brain emerges far greater than the sum of its parts. We still don't understand how thoughts or personality or behaviours are seeded and take root in this network, or how the neurons become organized to drive such processes, but somehow consciousness is created from the most prosaic building materials. We can describe the intelligence, creativity and sociability of Homni's collective brain as comparable to the networked, linked-up, conversational accumulation of all the billions of human brains, including those from the past who have left a cultural and intellectual legacy, and also the artificial brains of our technological inventions such as computer programs. Homni's global empire is presided over by multinational corporations, we communicate over global social media platforms and trade globally with the American dollar. We all log onto the same Internet,[14] eat pasta, pizza and rice in every city, buy jeans, drink cola, chew gum while listening to pop music. Despite their manifold inefficiencies, Homni exerts a global political authority and judicial system through the United Nations, whilst the trade of commodities between nations is governed by World Trade Organization rules and health care by the

World Health Organization. For many, the meaning of family, and of tribe and of nation, has dwindled in the face of belonging to our global networked society, and people are beginning to identify as global rather than national citizens.[15]

Currently there is no significant biological difference between people of different cultures – genetic features overlap and are dispersed across cultural and geographical boundaries, so that genetic variation within our populations is at least as great as between them. But in the future that might not be so: different cultural phenotypes could ascribe real biological differences, and today's adaptive human that can become any cultural phenotype would be no more. In coming decades, individuals who remain outside of our superorganism may find themselves isolated culturally, technologically and even physically and cognitively. Describing a human will increasingly assume an increased life span and expansive communicative abilities, for instance. To fall outside of this new norm will be to belong to a biologically different race of human, perhaps even a subspecies. That is not because any of our cultures is superior or 'more evolved' than any other – relying on technological complexity does not necessarily make life more joyful or more meaningful than the lifestyles of hunter-gatherer societies (many would argue it is the reverse). Yet such societies are being squeezed out by the industrial lifestyles of vast populations of networked people whose command of energy is better. As Homni becomes ever more homogenized,[16] we would do well to remember the importance of maintaining diversity in our cultural and biological toolbox; it is a survival adaptation that has served us well in the past and could prove invaluable as we enter uncharted territory. That means protecting the rights of all humans – and the places where they live – from our marauding superorganism.

For Homni also exerts a strong physical presence. While individual humans and societies will have local or regional environmental effects, our superorganism has made changes on such a scale that they have altered our world beyond anything it has experienced in its 4.6-billion-year history. Our planet is crossing another geological boundary, and this time we are the change makers. Geologists are calling this new epoch the Anthropocene (meaning the Age of Humans), recognizing that humanity has become a geophysical force

on a par with Earth-shattering asteroids and planet-cloaking volcanoes that defined past epochs. The environment that shaped human
evolution has been fundamentally changed by us.

In a single lifetime we've become a phenomenal global force and
there is no sign of a slowdown. Two-fifths of land surface is used to
grow our food; we control three-quarters of the world's freshwater;
no part of the planet is untouched by us – we even decide the temperature of the atmosphere. We have gone from an endangered puny
primate on the savannahs of Africa to become the most numerous big
animal on Earth, and the next in line are the animals we have created
through breeding to feed and serve us. Our voracious plundering of
the natural world has led to massive deforestation and a surge in
extinctions and destroyed ecosystems. It will take millions of years
for mammals to recover the evolutionary diversity that humanity has
destroyed (more than ten times longer than we have existed). We have
also produced a deluge of waste that will take centuries to degrade.
When we hunt wild fish from the oceans, we eat with them the plastic
garbage that we discard. We have shrunk the endless landscapes of
our planet to within the operating scale of our own activities. And
the consequences of our Anthropocene will be faced for generations
to come: we have colonized the future.[17]

Our cultural evolution has given Homni the power to dramatically
shift the fortunes of every species, including our own. However, it is
our position within Homni's 'connectome' – our collective brain
network – that dictates our individual lives to a greater extent than our
biology or genetics. In the same city, a person born white, male, from
a prestigious land-owning family in a wealthy Western nation will lead
a different life to a dark-skinned refugee from the global south without
status or wealth.[18] Their IQ, physical and mental health, fitness, political persuasion, diseases, number of children, future wealth and life
expectancy are all strongly influenced by their connections. And these
differences are likely to be culturally inherited by at least one generation. When the slime mould coalesces, some amoeba cells are nicely
protected in the heart of the superorganism, whereas others find themselves weakly connected on the outside where they are vulnerable.

The human evolutionary triad – genes, environment and culture –
all affect the way the network is shaped, and this determines how we

operate as a society. As Homni blunders along, our illusion of free will is just that. And yet we cling to it because in spite of Homni's dominion, we can each of us influence others through this network and, therefore, potentially influence the beast itself. The most remarkable thing about our superorganism is that, unlike amoebae, it is made up of billions of unrelated individuals. Homni has emerged uniquely from natural evolutionary processes.

From an evolutionary perspective, the meaning of life is to perpetuate our genes, and during our ancestry we developed such a successful way of doing this through culture that we now dominate all life on Earth. Yet our cultural purpose – self-determination – has eclipsed that of our biology: we have power to pick our genes, to decide who lives and who dies, even to extinguish our entire species. If we are to survive, then our cultural evolution will have to take the next step to go from acting on group survival to global population survival: Homni's survival.

Perhaps the biggest lesson for the Anthropocene – of our increasing self-awareness as a planetary species – is that the rules for cultural evolution apply equally to the biological evolution of our environments: we need to maintain species populations and connectivity if we are going to see ecological diversity and complexity. While the size of Homni's network brings him benefits from increasing returns in technological and cultural complexity and diversity, this is increasingly at the expense of our environment. Earth's resources are not infinite and Homni is already using a quarter of the planet's net primary production.[19] This is unsustainable, and will deliver us diminishing returns. Deciding, though, as an individual, to reduce freshwater waste or to cut a carbon footprint, has a negligible impact. Although individuals may be able to steer Homni to some degree, it is far from obvious how we might meet the planetary challenges of the Anthropocene, and yet our changed planet will itself strongly influence Homni.[20]

The Anthropocene is likely to be just as culturally transformative as our last geological transition from the Pleistocene to the Holocene – but while that event took place over millennia, this is occurring within decades. We may see metres of sea-level rise in the lifetime of my children, which could devastate our human world and possibly our civilization. A change in temperature of just a degree or so has produced enormous social upheaval for past cultures such as the Roman and Mayan empires, and we've already seen a similar temperature rise in the Anthropocene, contributing to wars, regional instability and several million refugees. Our cultures will need to adapt as never before to the new world we are creating.[21]

Our biology is already changing: sperm counts among men in the West have more than halved;[22] and we are malnourished in new ways, despite an obesity epidemic[23] affecting more than one-third of adults. It's amazing to think that we now devote entire industries to getting fewer calories from food, when hundreds of thousands of years of evolution were devoted to the precise opposite.[24]

We are continuing to evolve, driven by our technologies and social norms: our foreheads are expanding, we are growing taller and developing a much higher incidence of myopia,[25] for instance. These changes are happening slowly, for Darwinian evolution usually operates at a slower pace than cultural. However, we are acquiring the skills to speed up our genetic evolution, through vaccination or by using tools that hack our DNA directly during IVF. The latest gene-editing technology, CRISPR, invented in 2012, acts like molecular scissors, snipping out particular genes and inserting other bits into the genome. It allows rapid, easy, and accurate editing of life's blueprint, and its potential is enormous. Today, we have the ability to create novel life forms, from new crop varieties to new human beings, one gene-switch at a time. It is already possible to eliminate genes that cause some fatal conditions; one day it may be possible to defeat death itself. Meanwhile, personalized treatments tailored to our genetic and biological profile, and lab-grown organs, tissues and cells, will improve life expectancy.

As we continue to improve our natural abilities with artificial body parts, cyborgs like Neil Harbisson will become more common. Our blood and organs may host nanobots that monitor our health and deliver targeted drugs. Humans will become increasingly designed.

As Homni evolves, the biological proportion of our superorganism will be augmented by a growing robotic component – we already share the planet with 9 million robots and our collective brain increasingly includes artificial intelligence as we outsource not just the energy requirements of our big brains, but also the big brains themselves. We are heavily reliant on artificial memory and processing: Homni's annual data footprint is already equivalent to 40 sextillion[26] bits, or around 5 zettabytes a year – an unimaginable number of zeros and ones. We are culturally evolving more and more crutches to offload our cognitive activities, through our computing and social resources, which may be making us more stupid. Thousands of years ago, Socrates worried about the effect of another technology – writing – saying it would erode young people's ability to remember.[27] He was right; rote memorizing is redundant, but we may be getting better at other tasks – such as dealing with abstract information – because our cultural development bath in the industrialized world equips us to see patterns and think in symbols and categories from a young age. This has led to an average rise in IQ of 30 points in the past 80 years (the so-called Flynn effect), whereas our navigation skills have deteriorated.[28]

AI is perhaps the ultimate manifestation of our brain's evolved desire for predictability. Little is more predictable than a human-designed algorithm. At many repetitive tasks, computer programs have already proved themselves to be far better than humans, and the goal is for machines to be able to learn how to perform tasks and make decisions independently. This makes AI perfect for jobs that involve large datasets of information, where statistical outcome is more important than subjective values, and they are often faster and more accurate than humans, since it takes us longer to remember or look up information and we are prone to biases, fatigue and boredom.

But what happens when AI make a mistake? At the moment, our social norms make allowances for some human error, but we expect our machine decisions to be 100 per cent accurate all the time. There is already a long list of AIs that make bad decisions because of coding mistakes or biases in the data they trained on, making AI worryingly human but less accountable. Another issue is privacy – to optimize AI, you need to supply it with the most comprehensive dataset. Increasingly, our data – which are essentially our reputations – are being controlled by a few multinational corporations, and could potentially be used against us. Genome testing companies collect masses of personal data, and genealogy databases can already identify

60 per cent of Americans, even if they haven't been tested. Big data-sets will make Homni a far more efficient planetary player, but if our reputations are not safeguarded, we risk individual tragedies and greater social inequalities. States today have unprecedented access to the private lives of citizens – in China, this has enabled the government to develop a credit-score algorithm that spies on people and uses data about their behaviours and friendships to rank their 'cultural worth'. People with low scores are blacklisted and denied plane tickets, jobs and credit.

These are all real and significant issues with AI, but manageable with good governance. Artificial intelligence, for all its promise and threat, is not about to replace humanity. We are more able, flexible and multi-talented than the most advanced robots. Calculating and manipulating data, although impressive, is not the pinnacle of human intelligence – indeed, someone with these abilities but lacking any common sense and social nous would be diagnosed with cognitive disorder. However, there's no doubt that many human roles will increasingly be taken by robots – they are simply more efficient and, as we have seen, energy efficiency is the underlying driver of our cultural evolution. The problem is, humans (unlike robots) derive a sense of purpose, identity and meaning from work, and without social planning, we risk an unstable and inhumane transition into this next economy.

I started working on this book with a vague understanding of our human story as a progression: from the miserable hardships of ape-men to the happy citizens enjoying the comforts and conveniences of the modern world. It's astonishing to note that despite thousands of years of technological progression, it has only been in last few centuries that we've experienced any real improvement in human welfare. By several measures, things are now better than they've ever been: In 1500, people in London were no better off than those in Delhi; now, there is no country in the world that has a child mortality rate as high as Portugal's was in 1950. From the 1800s, well-being for ordinary people rose significantly, mostly due to scientific advances in agriculture and medicines.[29] Today we have the safest, most plentiful and most affordable food supply in all of human history.

We still fight each other,[30] but the proportion of the population dying in wars has fallen. (This is not necessarily because we have become less violent, but because there is safety in numbers – the same is seen with primates.) Homni makes global wars less likely, partly because of the nuclear threat, but mainly because we are all so interconnected and interdependent now through our economies, trade, family and cultural practices. Homni's world is a safer world for humans and a better one. But there is nothing inevitable about our continued progression.

When I look at the news, I see the same social problems we've battled for millennia, such as tribalism, and the eternal tension between individual self-interest and the collective good. I've seen the United Kingdom riven along partisan lines as it attempts to leave the greatest peaceful collaboration in history; the rise of fascism in liberal democracies; hate-filled rhetoric by the president of the United States against citizens of a different gender and race; millions fleeing war and violence in Africa, Asia and the Middle East; and global inaction towards preventing environmental catastrophe. For all of our technological progress, socially we are in many ways regressing – the norms that allow large multicultural societies to live together harmoniously and productively are breaking down. Inequalities between groups mean their interests aren't necessarily aligned; they don't identify as belonging to the same tribe, and this builds conflict over cooperation. Although our technologies ratchet up in sophistication, we seem unable to stop repeating the same social mistakes of the past, as though there is a fault in humanity's cultural algorithm.

Truly, there are reasons for pessimism and despair, but much of this is a problem of perspective. We are compelled to live only in our time, and so the minutiae of social and political life are for us epic dramas. Yet on the perspective of human cultural evolution, our little lives are mere ripples on an ocean swell of change – humans may regress to a dark age of racial inequality before riding a wave of improved human rights. I wonder if these peaks and troughs are in fact part of a greater upswell; we may be heading somewhere grander and better. In bleak times, it can be helpful to remember the many kindnesses of humans and the courage of individuals that enable us to make great social improvements in a very brief time. The abolition of slavery, the introduction of rights for women and universal health care are all things that once seemed unthinkable, yet were spearheaded by a few individuals and changed the lives of millions. Homni is a formidable force, because he is made up of billions of humans who are, individually, remarkable. More than a quarter of the world's population are children, still acquiring the cultural knowledge they will need to help solve humanity's biggest challenges. They will develop new technologies, new social norms, new ways of finding meaning and new ways of interacting with the natural world. But they will only realize their enormous human potential if they are nurtured in a kind, collaborative and inclusive developing bath, because even though we operate on the planetary scale as part of Homni, we still live in communities of a few hundred people. Only by recognizing and embracing our shared humanity on our one living planet will we achieve a good, livable Anthropocene.

We are now at a point of unprecedented genetic, cultural and environmental power as a species, and we are linked to virtually every other person on Earth. We are embodied individuals trapped in a temporal existence, but we are also networked data streams, memories and influencers, and part of a grander humanity. Our decisions today have far-reaching consequences that imbue us with a responsibility to become good ancestors, to take the long view and time-travel forward to imagine the well-being of billions of people whose lives will be lived in the world we are currently making. Centuries ago, leaders of the indigenous North American Iroquois people created 'seven generation stewardship', instructing people to consider the impact of every decision on their children, seven generations into the future. In the precious few decades that Earth is ours, while we enjoy the gardens planted by our ancestors, we must not steal the shade from our descendants.

As I write this, high in the night sky a very constant shooting star has crossed my window. It is the International Space Station, an extraterrestrial home in space permanently occupied by the only life form capable of doing so. Through hundreds of thousands of years of human collaboration we have achieved the most incredible magic. We are all part of something extraordinary: our iterations of the body of our collective culture take us in unpredictable directions. They create for us new problems but also, we hope, their solutions. After all, there is nobody but us.

Acknowledgements

Like every cultural product, this book is born of the collective efforts of millions of people past and present, and I'm particularly grateful to those who wrote books, created libraries and museums, and made so much information accessible to me. I've gained so much from the generosity of countless individuals and communities I've visited in my travels around the world. The trite observation that we are all different and yet all the same is nonetheless intriguing, and it's been fascinating to explore.

I thank the many scientists and others who shared their specialist knowledge with me during this research, including Maggie Boden, Mark Thomas, Mark Pagel, Eske and Rane Willeslev, Chris Stringer, Robert Boyd, Nicholas Christakis, Clive and Geraldine Finlayson, Panos Athanasopoulos, Thomas Bak, Jubin Abutalebi, David Rand and Molly Crockett.

I am so grateful to Helen Conford and T. J. Kelleher, both of whom took a punt on my proposal for this book and bought it for Allen Lane and Basic Books. And to everyone who has given their time towards its creation, including my careful copy editor Bill Warhop and great editorial and publicity teams, including Laura Stickney, Michelle Welsh-Horst, Holly Hunter, Richard Duguid, Annabel Huxley and Julie Woon.

None of this would have been possible without my wonderful friend and agent Patrick Walsh, who has supported me through my various crises and managed to make me see the funny side. On that note, I am grateful to Marina Hyde, Ian Dunt, John Crace, @ManWhoHasItAll and other writers for their light during bleak times. And to my kind and stoic friends who have supported me in their own ways, including Jolyon Goddard, Helen Czerski, John Ash, Deborah Cohen, Michael

Regnier, Michelle Martin, Sara Abdulla, John Witfield, Charlotte and Henry Nicholls, Mesi Ashebir, Olive Heffernan, Rowan Hooper, Kat Mansoor, Brian Hill and my Sisters in Arms: Jo Marchant and Emma Young, who have kept me just this side of sanity.

Somewhere during the all-consuming process of writing it, my book became known as the Poopoo Book for stealing my family time. Heartfelt thanks to my much more important babies Kipp and Juno, my parents Ivan and Gina, and my partner Nick for all your love and support.

Notes

INTRODUCTION

1. Christakis, N., and Fowler, J. Friendship and natural selection. *Proceedings of the National Academy of Sciences* 111, 10796–10801 (2014).
2. Boardman, J., Domingue, B., and Fletcher, J. How social and genetic factors predict friendship networks. *Proceedings of the National Academy of Sciences* 109, 17377–17381 (2012).

1. CONCEPTION

1. Which changes Earth's orbit in regular 405,000-year cycles, affecting climate. Kent, D., et al. Empirical evidence for stability of the 405-kiloyear Jupiter–Venus eccentricity cycle over hundreds of millions of years. *Proceedings of the National Academy of Sciences* 115, 6153–6158 (2018).
2. Wolfe, J. Palaeobotanical evidence for a June 'impact winter' at the Cretaceous/Tertiary boundary. *Nature* 352, 420–423 (1991).
3. There's a wonderful description of this event in Brannen, P. *The ends of the world* (HarperCollins, 2017).
4. DeCasien, A., Williams, S., and Higham, J. Primate brain size is predicted by diet but not sociality. *Nature Ecology & Evolution* 1 (2017).

2. BIRTH

1. Much of the world's water was locked up in ice sheets.
2. According to a new predictive model, human brain size evolved in response to a number of different factors that were 60 per cent ecological, 30 per cent cooperation-related, and 10 per cent related to competition between groups. González-Forero, M., and Gardner, A. Inference of

ecological and social drivers of human brain-size evolution. *Nature* 557, 554–557 (2018).

3. Huff, C., Xing, J., Rogers, A., Witherspoon, D., and Jorde, L. Mobile elements reveal small population size in the ancient ancestors of Homo sapiens. *Proceedings of the National Academy of Sciences* 107, 2147–2152 (2010).

4. Hublin, J., et al. New fossils from Jebel Irhoud, Morocco and the pan-African origin of Homo sapiens. *Nature* 546, 289–292 (2017).

5. Modern humans did, it seems, journey out of Africa around 180,000 years ago, and on a few other occasions interbreed with the people they met, but these humans were unsuccessful and it would take another 100,000 years before our emigration succeeded to leave surviving descendants: us. There are clues that at least some modern humans may have departed Africa well before 100,000 years ago, only to go locally extinct. Harvati, K., et al. Apidama cave fossils provide earliest evidence of *Homo sapiens* in Eurasia. *Nature* (2019) doi.org/10.1038/s41586-019-1376-2 In Papua New Guinea, some 2 per cent of people's DNA seems to be from much older modern humans. They may carry a trace of DNA from an earlier wave of Africans who left the continent as long as 140,000 years ago, and then vanished.

6. And they bred with each other, as the child of a Neanderthal mother and Denisovan father reveals: Slon, V., et al. The genome of the off-spring of a Neanderthal mother and a Denisovan father. *Nature* 561, 113–116 (2018).

7. Lachance, J., et al. Evolutionary history and adaptation from high-coverage whole-genome sequences of diverse African hunter-gatherers. *Cell* 150, 457–469 (2012).

8. The entire Neanderthal population had less genetic diversity than any living human group.

9. Nielson, R., et al. Tracing the peopling of the world through genomics. *Nature* 541, 302–310 (2017).

3. LANDSCAPING

1. McPherron, S., et al. Evidence for stone-tool-assisted consumption of animal tissues before 3.39 million years ago at Dikika, Ethiopia. *Nature* 466, 857–860 (2010).

2. It is thought that the phase of dramatic shifts in climate and vegetation that occurred around 2 million years ago spurred our ancestors to evolve our adaptive versatility to different environments.

3. Gowlett, J., and Wrangham, R. Earliest fire in Africa: Towards the convergence of archaeological evidence and the cooking hypothesis. *Azania: Archaeological Research in Africa* 48, 5–30 (2013).

4. Heyes, P., et al. Selection and use of manganese dioxide by Neanderthals. *Scientific Reports* 6 (2016).

5. Some evolutionary anthropologists argue we have a 'fire-learning instinct', because children show the same inbuilt response to fire as they do toward predatory animals, including being motivated to pay special attention to information about it. Fessler, D. A burning desire: Steps toward an evolutionary psychology of fire learning. *Journal of Cognition and Culture* 6, 429–451 (2006).

6. Domínguez-Rodrigo, M., et al. Earliest porotic hyperostosis on a 1.5-million-year-old hominin, Olduvai Gorge, Tanzania. *PLoS ONE* 7, e46414 (2012).

7. Perkins, S. Baseball players reveal how humans evolved to throw so well. *Nature* (2013). doi:10.1038/nature.2013.13281.

8. Geneticists have discovered a gene that controls both body hair and sweat gland production, and they appear to be inversely linked. Experiments in mice revealed that when the gene is activated, mice develop more sweat glands than hair; when it is inactive, they produce more hair than sweat.

9. I now 'hunt' and gather from my sofa using an online supermarket, obtaining the same calories as my distant ancestors through the physically exerting taps of my index finger.

10. Domínguez-Rodrigo, M., et al. Earliest porotic hyperostosis on a 1.5-million-year-old hominin, Olduvai Gorge, Tanzania. *PLoS ONE* 7, e46414 (2012).

11. Sakai, S., Arsznov, B., Hristova, A., Yoon, E., and Lundrigan, B. Big cat coalitions: A comparative analysis of regional brain volumes in Felidae. *Frontiers in Neuroanatomy* 10 (2016).

12. Daura-Jorge, F., Cantor, M., Ingram, S., Lusseau, D., and Simoes-Lopes, P. The structure of a bottlenose dolphin society is coupled to a unique foraging cooperation with artisanal fishermen. *Biology Letters* 8, 702–705 (2012).

13. Henrich, J. *The secret of our success* (Princeton University Press, 2015).

14. Hung, L., et al. Gating of social reward by oxytocin in the ventral tegmental area. *Science* 357, 1406–1411 (2017).

15. Some had speculated that climate change may have caused the drop, but the numbers of smaller carnivores, which are more sensitive to climate anomalies, showed no such decline, rather fingering our role.

16. Fires also played an important role in communication. Smoke can be seen from tens of kilometres away if the weather conditions are right, transmitting farther and more persistently than sound. Smoke generated by placing damp grass or leaves on a fire allows people to signal information to faraway members of the group, or to inform neighbouring tribes of a peaceful approach to their territory.

4. BRAIN BUILDING

1. Kentucky midwife delivers her own baby via C-section. *Lexington Herald Leader* (2018). https://www.kentucky.com/news/state/article 205079969.html.

2. Fox, K., Muthukrishna, M., and Shultz, S. The social and cultural roots of whale and dolphin brains. *Nature Ecology & Evolution* 1, 1699–1705 (2017).

3. Only an ant and marmoset have bigger brains relative to body size.

4. Suzuki, I., et al. Human-specific NOTCH2NL genes expand cortical neurogenesis through delta/notch regulation. *Cell* 173, 1370–1384.e16 (2018).

5. Caceres, M., et al. Elevated gene expression levels distinguish human from non-human primate brains. *Proceedings of the National Academy of Sciences* 100, 13030–13035 (2003).

6. Burgaleta, M., Johnson, W., Waber, D., Colom, R., and Karama, S. Cognitive ability changes and dynamics of cortical thickness development in healthy children and adolescents. *NeuroImage* 84, 810–819 (2014).

7. Powell, J., Lewis, P., Roberts, N., Garcia-Finana, M., and Dunbar, R. Orbital prefrontal cortex volume predicts social network size: An imaging study of individual differences in humans. *Proceedings of the Royal Society B: Biological Sciences* 279, 2157–2162 (2012).

8. Tamnes, C., et al. Brain maturation in adolescence and young adulthood: Regional age-related changes in cortical thickness and white matter volume and microstructure. *Cerebral Cortex* 20, 534–548 (2009).

9. Kaplan, H., and Gurven, M. *The natural history of human food sharing and cooperation: A review and a new multi-individual approach to the negotiation of norms* (MIT Press, 2005).

10. Tronick, E., Morelli, G., and Winn, S. Multiple caretaking of Efe (Pygmy) infants. *American Anthropologist* 89, 96–106 (1987).

11. They derive indirect fitness benefits – impacts that individuals have on the reproduction and survival of their relatives – through cooperating.

Dyble, M., Gardner, A., Vinicius, L., and Migliano, A. Inclusive fitness for in-laws. *Biology Letters* 14, 20180515 (2018).

12. Hamlin, J. The case for social evaluation in preverbal infants: Gazing toward one's goal drives infants' preferences for Helpers over Hinderers in the hill paradigm. *Frontiers in Psychology* 5 (2015).

13. Hamlin, J., Wynn, K., Bloom, P., and Mahajan, N. How infants and toddlers react to antisocial others. *Proceedings of the National Academy of Sciences* 108, 19931–19936 (2011).

14. Noss, A., and Hewlett, B. The contexts of female hunting in central Africa. *American Anthropologist* 103, 1024–1040 (2001).

15. The grandmother effect has also been observed in whales, with grandmother orcas food sharing with their children and grandchildren well into adulthood.

16. Hawkes, K., O'Connell, J., and Blurton Jones, N. Hadza women's time allocation, offspring provisioning and the evolution of long post-menopausal life spans. *Current Anthropology* 38, 551–577 (1997).

17. In the United States, this has led to a culture where one in four mothers returns to work less than two weeks after giving birth – many working long shifts of up to 12 hours just a week after birth.

18. Hrdy, S. *Mother nature* (Ballantine Books, 2000).

19. Fonseca-Azevedo, K., and Herculano-Houzel, S. Metabolic constraint imposes tradeoff between body size and number of brain neurons in human evolution. *Proceedings of the National Academy of Sciences* 109, 18571–18576 (2012).

20. Although one study estimates our number of neurons at just 86 billion. Herculano-Houzel, S. The human brain in numbers: A linearly scaled-up primate brain. *Frontiers in Human Neuroscience* 3 (2009).

21. And these genetic changes were vital – people with mutations in these genes are unable to get sufficient glucose across the blood-brain barrier, causing learning disabilities, seizures or microcephaly – a miniaturization of the brain.

22. Churchill, S.E. Bioenergetic perspectives on Neanderthal thermoregulatory and activity budgets. In *Neanderthals revisited: New approaches and perspectives. Vertebrate paleobiology and paleoanthropology.* Ed. Hublin, J. J., Harvati, K., and Harrison, T. (Springer, 2006).

23. Wrangham, R. *Catching fire* (Profile, 2009).

24. Cordain, L., et al. Plant-animal subsistence ratios and macronutrient energy estimations in worldwide hunter-gatherer diets. *American Journal of Clinical Nutrition* 71, 682–692 (2000).

25. Lamichhaney, S., et al. Rapid hybrid speciation in Darwin's finches. *Science* 359, 224–228 (2017).

26. Bibi, F., and Kiessling, W. Continuous evolutionary change in Plio-Pleistocene mammals of eastern Africa. *Proceedings of the National Academy of Sciences* 112, 10623–10628 (2015).

27. It seems that the opposite happened to the ancestors of the blind cave fish of Mexico. On finding themselves trapped for millennia in a dark cave, their brains shrank dramatically and sacrificed redundant vision in order to save energy and ensure their survival.

28. Mitteroecker, P., Windhager, S., and Pavlicev, M. Cliff-edge model predicts intergenerational predisposition to dystocia and Caesarean delivery. *Proceedings of the National Academy of Sciences* 114, 11669–11672 (2017).

29. Shatz, S. IQ and fertility: A cross-national study. *Intelligence* 36, 109–111 (2008).

30. Juvenal:
    What a monstrous maw that feeds
    On a whole wild boar, a creature that's fit for a banquet!
    There's swift punishment though, when bloated you doff
    Your cloak, and go for a bath, with a part-digested peacock
    Inside. Then for the old it's death, intestate and sudden.
    The news is passed round at dinner, with never a tear;
    And the funeral's performed to the cheers of irate friends.

31. Pliny the elder:
    These practices have ruined the morals of the Empire, I mean the practices to which we submit when in health ... boiling baths, by which they have persuaded us that food is cooked in our bodies, so that everyone leaves them the weaker for treatment, and the most submissive are carried out to be buried.

5. CULTURAL LEVERS

1. Gómez-Robles, A., Hopkins, W., Schapiro, S., and Sherwood, C. Relaxed genetic control of cortical organization in human brains compared with chimpanzees. *Proceedings of the National Academy of Sciences* 112, 14799–14804 (2015).

2. Enquist, M., Strimling, P., Eriksson, K., Laland, K., and Sjostrand, J. One cultural parent makes no culture. *Animal Behaviour* 79, 1353–1362 (2010).

3. Lewis, H., and Laland, K. Transmission fidelity is the key to the build-up of cumulative culture. *Philosophical Transactions of the Royal Society B: Biological Sciences* 367, 2171–2180 (2012).

4. Simonton, D. K. Creativity as blind variation and selective retention: Is the creative process Darwinian? *Psychological Inquiry* 10, 309–328 (1999).

5. Henrich, J., and Boyd, R. On modeling cultural evolution: Why replicators are not necessary for cultural evolution. *Journal of Cognition and Culture* 2, 87–112 (2002).

6. Rendell, L., et al. Why copy others? Insights from the Social Learning Strategies Tournament. *Science* 328, 208–213 (2010).

7. Tomasello, M. The ontogeny of cultural learning. *Current Opinion in Psychology* 8, 1–4 (2016).

8. Morgan, T., et al. Experimental evidence for the co-evolution of hominin tool-making teaching and language. *Nature Communications* 6 (2015).

9. Deino, A., et al. Chronology of the Acheulean to Middle Stone Age transition in eastern Africa. *Science* 360, 95–98 (2018).

10. Munoz, S., Gajewski, K., and Peros, M. Synchronous environmental and cultural change in the prehistory of the northeastern United States. *Proceedings of the National Academy of Sciences* 107, 22008–22013 (2010).

11. Johnson, B. 65,000 years of vegetation change in central Australia and the Australian summer monsoon. *Science* 284, 1150–1152 (1999).

12. Klarreich, E. Biography of Richard G. Klein. *Proceedings of the National Academy of Sciences* 101, 5705–5707 (2004).

13. Wei, W., et al. A calibrated human Y-chromosomal phylogeny based on resequencing. *Genome Research* 23, 388–395 (2012).

14. Powell, A., Shennan, S., and Thomas, M. Late Pleistocene demography and the appearance of modern human behavior. *Science* 324, 1298–1301 (2009).

15. Collard, M., Buchanan, B., and O'Brien, M. Population size as an explanation for patterns in the Paleolithic archaeological record. *Current Anthropology* 54, S388–S396 (2013).

16. Wilkins, J., Schoville, B., Brown, K., and Chazan, M. Evidence for early hafted hunting technology. *Science* 338, 942–946 (2012).

17. Pottery arrives in Britain in 4000 BCE, contemporary with agriculture as a perfected technological process – there are no experimental examples of small trinkets in Britain, as with, say, metalworking. Pottery was an existing technology that suddenly became very useful for agriculture, where populations were settled and needed vessels. The link between pottery and development of agriculture is so great that archaeologists

use evidence of pottery to infer farming in areas where no evidence for agriculture exists. However, in the Near East archaeologists find a 'pre-pottery Neolithic' with a long gap between the emergence of agriculture and pottery there.

18. Calmettes, G., and Weiss, J. The emergence of egalitarianism in a model of early human societies. *Heliyon* 3, e00451 (2017).

19. Making a ceramic vessel requires knowledge of how and where to obtain suitable fuels, clays and tempers (the 'grog' such as crushed sea-shells, fibres or gravel that prevent shrinkage in the clay and help it withstand thermal shock), as well as skills in the handling, mixing, shaping, drying, decorating and firing of the vessel.

20. People first began making things from metal around 11,000 years ago, when they used gold and copper – metals that exist in their native form in Earth's crust.

21. Miodownik, M. *Stuff matters: The strange stories of the marvellous materials that shape our man-made world* (Viking, 2013).

22. Discovering bronze was an amazing coup. When bronze was first made, it was with an alloy of arsenic rather than tin. But arsenic fumes are toxic and caused blacksmiths a variety of horrible illnesses and early death. Hephaestus, the Greek god of fire, volcanoes and metals, is usually shown limping and crippled – the buffoon of the other gods – probably because he suffered from arsenic poisoning. Later ancient Greek depictions of him show him healthy, perhaps because contemporary metal-workers had switched to tin alloys.

23. Metals are flexible and can be reshaped because they are made of crystals arranged in layers that slide over each other – the hotter they become, the more loosely the layers are held together. Alloys disrupt this because the crystals are no longer all made of the same metal atoms: some are replaced by alloy atoms, which change the regular properties of the layers and stop them sliding over each other so easily. The result is a much stronger metal.

24. The expense of bronze meant that civilizations at the time were highly aristocratic societies of narrow classes of nobles and priests ruling over masses of peasants still using stone tools.

25. The carbon doesn't replace the iron atoms in steel, as the tin does in bronze; rather it squeezes in alongside the iron atoms, changing the crystal.

26. 'The woods and groves are cut down, for there is need of an endless amount of wood for timbers, machines and the smelting of metals. And when the woods and groves are felled, then are exterminated the beasts and birds [. . .] Therefore the inhabitants of these regions, on account of the devastation of their fields, woods, groves, brooks and

rivers, find great difficulty in procuring the necessaries of life' (Georgius Agricola describing Bohemia in 1556).

6. STORY

1. Chatwin, B. *The songlines* (Franklin Press, 1987).
2. As we have seen, it is much more cost-effective in terms of time and energy to use the collective memory bank than for individual brains to hold and process vast amounts of cultural knowledge.
3. Bowdler, S. Human occupation of northern Australia by 65,000 years ago (Clarkson et al. 2017): A discussion. *Australian Archaeology* 83, 162–163 (2017).
4. In explaining how songlines work, Diana James wrote in her 2015 essay 'Tjukurpa Time' of how Nganyinytja, a Pitjantjatjara woman of elder high degree, learned to read her people's history written in the land: Long history, deep time: Deepening histories of place (2015). doi:10.22459/lhdt.05.2015.
5. In the end, it was colonization by gun-bearing Europeans that proved disastrous. Aboriginal land was taken, families were split, and the connection to their ancestors, which had been so crucial to survival, was severed. Within decades, indigenous communities lost their culture, language and lives.
6. Many people think of the Egyptians as the world's first bakers, with evidence of Egyptian bread dating from 17,000 BCE. But in fact Australian Aborigines were baking ngardu at least 30,000 years ago. And there is evidence of bread made from wild cereals from 14,500 years ago in Jordan Arranz-Otaegui, A., Gonzalez Carretero, L., Ramsey, M., Fuller, D., and Richter, T. Archaeobotanical evidence reveals the origins of bread 14,400 years ago in northeastern Jordan. *Proceedings of the National Academy of Sciences* 115, 7925–7930 (2018).
7. Romney, J. Herodotean geography (4.36–45): A Persian *Oikoumenē?* *Greek, Roman, and Byzantine Studies* 57, 862–881 (2017).
8. Bruner, J. *Actual minds, possible worlds* (Harvard University Press, 1987).
9. Stephens, G., Silbert, L., and Hasson, U. Speaker-listener neural coupling underlies successful communication. *Proceedings of the National Academy of Sciences* 107, 14425–14430 (2010).
10. The system is not foolproof, as we shall see later; however, for most of our physical interactions it works well enough.
11. Heider, F., and Simmel, M. An experimental study of apparent behavior. *American Journal of Psychology* 57, 243 (1944).

12. Seth, A. Consciousness: The last 50 years (and the next). *Brain and Neuroscience Advances* 2, 2398212818816o1 (2018).

13. Marchant, J. *Cure: A journey into the science of mind over body* (Canongate Books, 2016).

14. German doctors are six times more likely than American doctors to prescribe heart medication to patients exhibiting the same symptoms. Moerman, D., and Jonas, W. Deconstructing the placebo effect and finding the meaning response. *Annals of Internal Medicine* 136, 471 (2002).

15. Phillips D. P., Ruth T. E., and Wagner, L. M. Psychology and survival, *Lancet* 342, 1142–1145 (1993).

16. *World Health Organization Weekly Epidemiological Monitor* 5, 22 (2012). http://applications.emro.who.int/dsaf/epi/2012/Epi_Monitor_2012_5_22.pdf.

17. Kamen, C., et al. Anticipatory nausea and vomiting due to chemotherapy. *European Journal of Pharmacology* 722, 172–179 (2014).

18. In one clinical trial for antidepressants, a 26-year-old man who had overdosed on the tablets after taking 29 of them was rushed to hospital, suffering a plunging blood pressure of 80/40. He was being hooked up to intravenous fluids to stabilize him when clinicians revealed he had actually been taking placebo pills. His symptoms rapidly reversed on learning this: Reeves, R., Ladner, M., Hart, R., and Burke, R. Nocebo effects with antidepressant clinical drug trial placebos. *General Hospital Psychiatry* 29, 275–277 (2007).

19. Meador, C. Hex death. *Southern Medical Journal* 85, 244–247 (1992).

20. Ishiguro, K. The Nobel Prize in Literature 2017. *NobelPrize.org* (2019). https://www.nobelprize.org/prizes/literature/2017/ishiguro/25124-kazuo-ishiguro-nobel-lecture-2017/.

21. Bentzen, J. Acts of God? Religiosity and natural disasters across subnational world districts. *SSRN Electronic Journal* (2015). doi:10.2139/ssrn.2595511

22. Although believing in an angry, vengeful god can have the opposite effect, studies show.

23. Inzlicht, M., McGregor, I., Hirsh, J., and Nash, K. Neural markers of religious conviction. *Psychological Science* 20, 385–392 (2009).

24. Peoples, H., Duda, P., and Marlowe, F. Hunter-gatherers and the origins of religion. *Human Nature* 27, 261–282 (2016).

25. Smith, D., et al. Cooperation and the evolution of hunter-gatherer storytelling. *Nature Communications* 8 (2017).

26. Wiessner, P. Embers of society: Firelight talk among the Ju/'hoansi Bushmen. *Proceedings of the National Academy of Sciences* 111, 14027–14035 (2014).

27. Pearce, E., Launay, J., and Dunbar, R. The ice-breaker effect: Singing mediates fast social bonding. *Royal Society Open Science* 2, 150221 (2015).

28. The roots of this may go way back. Experiments show that apes who watch movies together feel socially closer subsequently. Wolf, W. and Tomasello, M. Visually attending to a video together facilitates great ape social closeness. *Proceedings of the Royal Society B* 286 (2019). doi.org/10.1098/rspb.2019.0488.

29. Pearce, E., et al. Singing together or apart: The effect of competitive and cooperative singing on social bonding within and between sub-groups of a university fraternity. *Psychology of Music* 44, 1255–1273 (2016).

30. Smith, D., et al. Cooperation and the evolution of hunter-gatherer storytelling. *Nature Communications* 8 (2017).

31. Dehghani, M., et al. Decoding the neural representation of story meanings across languages. *Human Brain Mapping* 38, 6096–6106 (2017).

32. Stansfield, J., Bunce, L. The relationship between empathy and reading fiction: Separate roles for cognitive and affective components. *Journal of European Psychology Students* 5, 9–18 (2014).

33. Kidd, D., and Castano, E. Reading literary fiction improves theory of mind. *Science* 342, 377–380 (2013).

34. Sala, I. What the world's fascination with a female-only Chinese script says about cultural appropriation. *Quartz* (2018). https://qz.com/1271372/.

35. Griswold, E. Landays: Poetry of Afghan women. *Poetry Magazine* (2018). https://static.poetryfoundation.org/o/media/landays.html.

36. da Silva, S., and Tehrani, J. Comparative phylogenetic analyses uncover the ancient roots of Indo-European folktales. *Royal Society Open Science* 3, 150645 (2016).

37. There are more than 700 of his persuasive morality tales, using animals to play out human dilemmas and storylines, often with a subversive message. In the strongly authoritarian times they were created, the fables regularly featured weak but clever animals succeeding over powerful individuals.

38. www.britishmuseum.org/research/collection_online/collection_object_details.aspx?objectId=176691&partId=1.

39. Dodds, E. *The Greeks and the irrational* (Beacon Press, 1957).

40. Mathews, R. H. Message-sticks used by the Aborigines of Australia. *American Anthropologist* 10, no. 9, 288–298 (1897).

41. Clayton, E. The evolution of the alphabet. British Library (2019). https://www.bl.uk/history-of-writing/articles/the-evolution-of-the-alphabet.

42. Kottke, J. Alphabet inheritance maps reveal its evolution clearly: The evolution of the alphabet. kottke.org (2019). https://kottke.org/19/01/the-evolution-of-the-alphabet.

43. Reading and writing have been lost in other places and times, too. After the Romans left Britain in 410 CE, for example, literacy skills were very nearly lost, and the British were saved only by Irish missionaries who recivilized the country and kept reading and writing alive until the Saxons arrived. But even then, it was a skill only achieved by a small section of society. Most people were illiterate.

44. It wasn't until the 700s BCE that the *Odyssey* and *Iliad* were written down.

45. It was not for another 500 years, when Phoenician traders introduced the alphabet, that they regained literacy, setting up schools and institutions to spread the skill.

46. Something that continued even into the twentieth century in parts of Europe where literacy was low.

47. Maguire, E., et al. Navigation-related structural change in the hippocampi of taxi drivers. *Proceedings of the National Academy of Sciences* 97, 4398–4403 (2000).

48. The technique was culturally transmitted across time and space, spreading to Rome and into the European Renaissance. Before printing, a trained memory was vital.

49. This isn't the case with Hebrew, which is very dense, with many anagrams.

50. Before the printing press, most literate Europeans used parchment (from animal skins) to write on, which only a small elite could afford as a single book required 250 sheep to produce the parchment.

51. DNA computers will use biochips holding logic gates made of DNA rather than silicon, making them cheaper, smaller, and yet capable of holding far more data. They can also process data faster because they can carry out calculations in parallel rather than purely linearly like conventional computers.

7. LANGUAGE

1. Wallace, E., et al. Is music enriching for group-housed captive chimpanzees (Pan troglodytes)? *PLoS ONE* 12, e0172672 (2017).

2. Back in the fifth century BCE, Herodotus described Ethiopian communities who 'speak like bats'.

3. Meyer, J. *Whistled languages: A worldwide inquiry on human whistled speech* (Springer-Verlag, 2015).

4. Wiley, R. Associations of song properties with habitats for territorial oscine birds of eastern North America. *American Naturalist* 138, 973–993 (1991).

5. Everett, C. Languages in drier climates use fewer vowels. *Frontiers in Psychology* 8 (2017).

6. Although any baby can currently learn any language, our genes may be evolving adaptations to make certain varieties easier to learn in different populations: Dediu, D., and Ladd, D. Linguistic tone is related to the population frequency of the adaptive haplogroups of two brain size genes, ASPM and Microcephalin. *Proceedings of the National Academy of Sciences* 104, 10944–10949 (2007).

7. Güntürkün, O., Güntürkün, M., and Hahn, C. Whistled Turkish alters language asymmetries. *Current Biology* 25, R706–R708 (2015).

8. Melody and rhythm of speech is very important, such as learning which are the strong and weak syllables in English.

9. Patel, A. Sharing and nonsharing of brain resources for language and music. *Language, Music, and the Brain* 329–356 (2013). doi:10.7551/mitpress/9780262018104.003.0014

10. Patel, A. Science and music: Talk of the tone. *Nature* 453, 726–727 (2008).

11. Blasi, D., et al. Human sound systems are shaped by post-Neolithic changes in bite configuration. *Science* 363, eaav3218 (2019).

12. Warner, B. Why do stars like Adele keep losing their voice? *Guardian* (10 August 2017).

13. Neanderthals, judging by their shorter vocal cords and wider nasal cavities, may have had higher-pitched voices and lacked the subtlety of human speech. Some think they wouldn't have been able to produce quantal vowels, which allow humans to distinguish between such words as 'beat' and 'bit', for example. Others disagree. Indeed, the evidence shows that the Neanderthals made the anatomical leap to enable language before *Homo sapiens,* so it's likely they were speaking long before our ancestors.

14. Although the jury's out on whether other 'language' genes were more important. Warren, M. Diverse genome study upends understanding of how language evolved. *Nature* (2018). doi:10.1038/d41586-018-05859-7.

15. Lai, C., Fisher, S., Hurst, J., Vargha-Khadem, F., and Monaco, A. A forkhead-domain gene is mutated in a severe speech and language disorder. *Nature* 413, 519–523 (2001).

16. Schreiweis, C., et al. Humanized FOXP2 accelerates learning by enhancing transitions from declarative to procedural performance. *Proceedings of the National Academy of Sciences* 111, 14253–14258 (2014).

17. Even a small child will, unasked, attempt to help an adult pick up something they have dropped, and this helpfulness is then shaped by the child's cultural developing bath – their social environment.

18. Russell, J., Gee, B., and Bullard, C. Why do young children hide by closing their eyes? Self-visibility and the developing concept of self. *Journal of Cognition and Development* 13, 550–576 (2012).

19. Moll, H., and Khalulyan, A. 'Not see, not hear, not speak': Preschoolers think they cannot perceive or address others without reciprocity. *Journal of Cognition and Development* 18, 152–162 (2016).

20. Partanen, E., et al. Learning-induced neural plasticity of speech processing before birth. *Proceedings of the National Academy of Sciences* 110, 15145–15150 (2013).

21. Hart, B., and Risley, T. The early catastrophe: The 30 million word gap by age 3. *American Educator* 27, 4–9 (2003).

22. Romeo, R., et al. Beyond the 30-million-word gap: Children's conversational exposure is associated with language-related brain function. *Psychological Science* 29, 700–710 (2018).

23. Researchers studying rhesus macaques off Puerto Rico have observed females interacting with babies using a special type of vocalization that appears to be a kind of motherese.

24. Brighton, H., and Kirby, S. Cultural selection for learnability: Three principles underlying the view that language adapts to be learnable. *Language Origins: Perspectives on Evolution* (2005). www.lel.ed.ac.uk/~kenny/publications/brighton_05_cultural.pdf.

25. Blasi, D., Wichmann, S., Hammarström, H., Stadler, P., and Christiansen, M. Sound–meaning association biases evidenced across thousands of languages. *Proceedings of the National Academy of Sciences* 113, 10818–10823 (2016).

26. Kirby, S. Culture and biology in the origins of linguistic structure. *Psychonomic Bulletin & Review* 24, 118–137 (2017).

27. Bromham, L., Hua, X., Fitzpatrick, T., and Greenhill, S. Rate of language evolution is affected by population size. *Proceedings of the National Academy of Sciences* 112, 2097–2102 (2015).

28. Orcas belonging to different groups also vocalize in different dialects.

29. Attributed to anthropologist Don Kulick, from Pagel, M. *Wired for culture: Origins of the human social mind* (W. W. Norton, 2012).

30. Speaking at HayFestival 2018.

31. In naming colours, most cultures start off with a prototype – for example, for the colour 'orange', the prototype is the orange fruit. After the 1500s, when the first orange trees were introduced to England, the fruit was such a distinctive colour that it was mentally registered as different from yellow and red (it was previously described as yellow-red). The word 'rayed' in Chaucer's work might have meant red in his time, but it might equally have meant orange or pink. 'Pink' used to mean not rose but yellow. In the fifteenth century, the English bought gallons of cheap paint called 'pink-yellow' (an impossibility now). The name probably originated from the German word 'pinkeln', meaning 'to tinkle' (urinate). The change in our meaning of the word 'pink' to the rose colour is likely to come from when we started to use pink-coloured distemper (as we still do) when undercoating walls, rather than yellow tinkle-coloured distemper in the late 1500s. The colours pink, purple and orange were only added in the past few hundred years.

32. Homer famously described the sea as 'wine-like'. Given that he was supposedly blind, this is understandable, however, there doesn't seem to be a single instance in ancient Greek literature of the sea or sky being simply 'blue'. This doesn't mean that the Greeks couldn't see different colours, just that the hue was only as important to them as, say, the brilliance and luminosity of the colour.

33. This may have been a more common linguistic tool in the past. In Scottish Gaelic, the words for up and down came to mean east and west or west and east, depending on which way the river closest to you flows to the sea. So in eastern Rosshire, people used to say: go east to the kitchen, even if the kitchen was west of where the person was standing, because it means go down to the kitchen in an area where downstream is in an easterly direction. In lots of languages, north, south, east and west are interchangeable with other meanings. For example, in English, we'll say things went south from there, meaning things went badly.

34. Boroditsky, L. How language shapes thought. *Scientific American* 304, 62–65 (2011).

35. Correia, J., Jansma, B., Hausfeld, L., Kikkert, S., and Bonte, M. EEG decoding of spoken words in bilingual listeners: From words to language invariant semantic-conceptual representations. *Frontiers in Psychology* 6 (2015).

36. Mårtensson, J., et al. Growth of language-related brain areas after foreign language learning. *NeuroImage* 63, 240–244 (2012).

37. According to one study, Speakers of 'future' languages – like English, which distinguishes past, present and future – are 30 per cent less likely

to save money than speakers of non-futured languages, like Chinese, perhaps because when the future is distinguished from the present, it appears more distant and so savers are less motivated.

38. Abutalebi, J., and Green, D. Control mechanisms in bilingual language production: Neural evidence from language switching studies. *Language and Cognitive Processes* 23, 557–582 (2008).

39. Being bilingual also delays dementia – in two people whose brains showed similar amounts of disease progression, the bilingual would show symptoms an average five years after the monolingual: Craik, F., Bialystok, E., and Freedman, M. Delaying the onset of Alzheimer disease: Bilingualism as a form of cognitive reserve. *Neurology* 75, 1726–1729 (2010). This is because bilingualism rewires the brain, boosting people's 'cognitive reserve'. It means that as parts of the brain succumb to damage, bilinguals can compensate more because they have extra grey matter and alternative neural pathways.

8. TELLING

1. Edemariam, A. The Saturday interview: Wikipedia's Jimmy Wales. *Guardian* (19 February 2019). https://www.theguardian.com/theguardian/2011/feb/19/interview-jimmy-wales-wikipedia.

2. Giles, J. Internet encyclopaedias go head to head. *Nature* 438, 900–901 (2005).

3. 'Wikipedian' was added to Oxford Dictionaries in 2012.

4. Nook, E., and Zaki, J. Social norms shift behavioral and neural responses to foods. *Journal of Cognitive Neuroscience* 27, 1412–1426 (2015).

5. Nook, E., Ong, D., Morelli, S., Mitchell, J., and Zaki, J. Prosocial conformity. *Personality and Social Psychology Bulletin* 42, 1045–1062 (2016).

6. Eriksson, K., Vartanova, I., Strimling, P., and Simpson, B. Generosity pays: Selfish people have fewer children and earn less money. *Journal of Personality and Social Psychology* (2018). doi:10.1037/pspp0000213.

7. From the perspective of each prisoner: if he doesn't testify against the other, he gets either one or three years in prison; whereas if he does testify, he goes free or gets two years in prison.

8. And this finding that individual cooperation depends on how generally cooperative the group is was also born out in a study of hunter-gatherers: Smith, K., Larroucau, T., Mabulla, I., and Apicella, C. Hunter-gatherers maintain assortativity in cooperation despite high levels of residential change and mixing. *Current Biology* 28, 3152–3157.e4 (2018).

9. Shirado, H., Fu, F., Fowler, J., and Christakis, N. Quality versus quantity of social ties in experimental cooperative networks. *Nature Communications* 4 (2013).

10. Crockett, M., Kurth-Nelson, Z., Siegel, J., Dayan, P., and Dolan, R. Harm to others outweighs harm to self in moral decision making. *Proceedings of the National Academy of Sciences* 111, 17320–17325 (2014).

11. Baillargeon, R., Scott, R., and He, Z. False-belief understanding in infants. *Trends in Cognitive Sciences* 14, 110–118 (2010).

12. Dunbar, R. Neocortex size as a constraint on group size in primates. *Journal of Human Evolution* 22, 469–493 (1992).

13. The first level of our Dunbar community is 'intimate friends' (we have an average of five of these); then 'best friends' (we have around 15); then 'bonded clan' (we have around 50 extended family and in-law relations), followed by our 'community' of 150 with whom we are in regular contact; 500 acquaintances and we recognize more than 1,500 faces.

14. Dunbar, R. Do online social media cut through the constraints that limit the size of offline social networks? *Royal Society Open Science* 3, 150292 (2016).

15. Jenkins, R., Dowsett, A., and Burton, A. How many faces do people know? *Proceedings of the Royal Society B: Biological Sciences* 285, 20181319 (2018).

16. Henrich, J., and Henrich, N. Culture, evolution and the puzzle of human cooperation. *Cognitive Systems Research* 7, 220–245 (2006).

17. Spies aim to discover wrongdoers when they do not know they are being watched, whereas police and surveillance systems rely on their visible presence to prevent criminals by showing them they are being watched. Large states use both systems to control their citizens.

18. Watts, J., et al. Broad supernatural punishment but not moralizing high gods precede the evolution of political complexity in Austronesia. *Proceedings of the Royal Society B: Biological Sciences* 282, 20142556–20142556 (2015).

19. Whitehouse, H., et al. Complex societies precede moralizing gods throughout world history. *Nature* 568, 226–229 (2019).

20. Lang, M., et al. Moralizing gods, impartiality and religious parochialism across 15 societies. *Proceedings of the Royal Society B: Biological Sciences* 286, 20190202 (2019).

21. Brennan, K., and London, A. Are religious people nice people? Religiosity, race, interview dynamics, and perceived cooperativeness. *Sociological Inquiry* 71, 129–144 (2001).

22. Chuah, S., Gächter, S., Hoffmann, R., and Tan, J. Religion, discrimination and trust across three cultures. *European Economic Review* 90, 280–301 (2016).

23. Embarrassment is closely related but needs an external observer.

24. Benedict, R. *The chrysanthemum and the sword. Patterns of Japanese culture* (Houghton Mifflin Harcourt, 1946).

25. Brass Eye. www.youtube.com/watch?v=f3xUjw2BCYE.

26. Cole, S., Kemeny, M., and Taylor, S. Social identity and physical health: Accelerated HIV progression in rejection-sensitive gay men. *Journal of Personality and Social Psychology* 72, 320–335 (1997).

27. Such cues fire up our brain's interest in learning and we find copying them more rewarding, brain imaging reveals.

28. Henrich, J., and Gil-White, F. The evolution of prestige: Freely conferred deference as a mechanism for enhancing the benefits of cultural transmission. *Evolution and Human Behavior* 22, 165–196 (2001).

29. Brands will pay prestigious people vast sums to be associated with their products. The flip side is that if someone loses reputation in one area, it impacts on their prestige in other areas, and brands are quick to cut ties with disgraced stars. Brands can now take out 'disgrace insurance' to safeguard themselves against celebrity ambassador mishaps, and the more squeaky clean a celebrity's image, the more expensive the insurance, because future disgrace is more likely to have an impact on their prestige, so brands have more to lose.

30. Hubris was particularly frowned on in ancient Greek times. It was one of the rare human foibles that the gods of *The Iliad*'s era got worked up about – and is a regular source of satirical material today.

9. BELONGING

1. Hunter, M., and Brown, D. Spatial contagion: Gardening along the street in residential neighborhoods. *Landscape and Urban Planning* 105, 407–416 (2012).

2. Asymmetries often arise from parasitic infections and are associated with developmental abnormalities.

3. Langlois, J., and Roggman, L. Attractive faces are only average. *Psychological Science* 1, 115–121 (1990).

4. Joshi, P., et al. Directional dominance on stature and cognition in diverse human populations. *Nature* 523, 459–462 (2015).

5. Lewis, M. Why are mixed-race people perceived as more attractive? *Perception* 39, 136–138 (2010).

6. Burley, N. Sex-ratio manipulation in colour-banded populations of zebra finches. *Evolution* 40, 1191 (1986).

7. Another theory for the rapid spread of light skin is that it may have offered a survival advantage. Evolutionary geneticist Mark Thomas points out that babies are all born lighter skinned than adults, in part because skin-darkening melanin develops in response to sun exposure. Children who retain this infantile paleness of skin and eyes elicit a protective response from adults, he says, and this puts them at a slight advantage over others. This intriguing idea can be extrapolated further: we have, after all, evolved to be hairless, domesticated, social creatures, with longer childhoods and playful childlike behaviour well into adulthood.

8. Ishizu, T., and Zeki, S. Toward a brain-based theory of beauty. *PLoS ONE* 6, e21852 (2011).

9. Little, A., Jones, B., and DeBruine, L. Facial attractiveness: Evolutionary based research. *Philosophical Transactions of the Royal Society B: Biological Sciences* 366, 1638–1659 (2011).

10. Brown, S., Gao, X., Tisdelle, L., Eickhoff, S., and Liotti, M. Naturalizing aesthetics: Brain areas for aesthetic appraisal across sensory modalities. *NeuroImage* 58, 250–258 (2011).

11. Jacobsen, T. Beauty and the brain: Culture, history and individual differences in aesthetic appreciation. *Journal of Anatomy* 216, 184–191 (2010).

12. The researchers think stick dolls are practice behaviour for the adult role of motherhood, because it is more common in females: Kahlenberg, S., and Wrangham, R. Sex differences in chimpanzees' use of sticks as play objects resemble those of children. *Current Biology* 20, R1067–R1068 (2010).

13. Dart, R. The water-worn Australopithecine pebble of many faces from Makapansgat. *South African Journal of Science* 70, 167–169 (1974).

14. Joordens, J., et al. Homo erectus at Trinil on Java used shells for tool production and engraving. *Nature* 518, 228–231 (2014).

15. d'Errico, F., Henshilwood, C., Vanhaeren, M., and van Niekerk, K. Nassarius kraussianus shell beads from Blombos Cave: Evidence for symbolic behaviour in the Middle Stone Age. *Journal of Human Evolution* 48, 3–24 (2005).

16. Social norms that regulate sexual fidelity may have evolved as an adaptation to larger societies because monogamy reduces sexually transmitted disease epidemics (and their fertility impacts), and helps ensure the children of a union are accepted as a father's responsibility for resource allocation.

17. Charles Darwin, when questioned on the matter, wrote: 'I certainly think that women though generally superior to men in moral qualities are inferior intellectually.' Darwin's theory was of his time but is untrue.

18. Chimpanzees live in quite aggressive, male-dominated societies with clear hierarchies. As a result, they just don't see enough adults in their lifetime for technologies to be sustained. Bonobos, by contrast, have fairly egalitarian social groups, far less conflict, and a range of collaborative cultures and behaviours – female bonobos form strong alliances with even unrelated females, and sometimes support each other during birth, for instance, including by swatting flies away.

19. Even women portraying women are influenced by this, as John Berger pointed out in his influential *Ways of seeing.*

20. Rothman, B. *The tentative pregnancy* (Pandora, 1988).

21. Some Pacific Islanders, for example, believe in the 'opacity' of other minds – the idea that it is impossible to know what other people think and feel. As a result, people are frequently held responsible for their wrongdoings, even when they were the result of an accident or error.

22. Centola, D., and Baronchelli, A. The spontaneous emergence of conventions: An experimental study of cultural evolution. *Proceedings of the National Academy of Sciences* 112, 1989–1994 (2015).

23. Touboul, J. The hipster effect: When anticonformists all look the same. *arXiv* (2014). http://arXiv:1410.8001v2.

24. Hipster whines at tech mag for using his pic to imply hipsters look the same, discovers pic was of an entirely different hipster. *The Register* (2019). http://www.theregister.co.uk/2019/03/06/hipsters_all_look_the_same_fact/.

25. Akdeniz, C., et al. Neuroimaging evidence for a role of neural social stress processing in ethnic minority–associated environmental risk. *JAMA Psychiatry* 71, 672 (2014).

26. As the literary theorist Kenneth Burke wrote: 'You persuade a man only insofar as you can talk his language by speech, gesture, tonality, order, image, attitude, identifying your ways with his.'

27. Hein, G., Silani, G., Preuschoff, K., Batson, C., and Singer, T. Neural responses to ingroup and outgroup members' suffering predict individual differences in costly helping. *Neuron* 68, 149–160 (2010).

28. Tribalism is such a strong driver, we also adapt our gods to suit our societies. For instance, Jesus is portrayed with blond hair and blue eyes in northern Europe, as black-skinned in Ethiopia and as indigenous Aymara in South America. Islam is unusual in that it discourages any form of portraiture and forbids showing the face of god or Muhammad.

This leads to bizarre compromises between depiction and non-depiction in Sunni-led Arabic countries, where on road signs, for example, a headless human figure will show pedestrians where to walk. Shia Islam is much more open to the depiction of people. In the fifteenth and sixteenth centuries, during the height of the colourful and detailed Persian miniatures movement, detailed portrayal of human figures, including Muhammad, were central to the genre under both Sunni and Shia rulers.

29. This effect is so strong that such tribal stories can remain powerful for hundreds of years and be called on to reliably stoke collective anger and hate. Thus, British Brexiters invoke the battles of Trafalgar and Agincourt, and, in the United States, white supremacists often display Confederate flags.

30. Dunham, Y., Baron, A., and Carey, S. Consequences of 'minimal' group affiliations in children. *Child Development* 82, 793–811 (2011).

31. Pope, S., Fagot, J., Meguerditchian, A., Washburn, D., and Hopkins, W. Enhanced cognitive flexibility in the seminomadic Himba. *Journal of Cross-Cultural Psychology* 50, 47–62 (2018).

32. Draganski, B., et al. Changes in grey matter induced by training. *Nature* 427, 311–312 (2004).

33. Gomez, J., Barnett, M., and Grill-Spector, K. Extensive childhood experience with Pokémon suggests eccentricity drives organization of visual cortex. *Nature Human Behaviour* (2019). doi:10.1038/s41562-019-0592-8.

34. Gislén, A., Warrant, E., Dacke, M., and Kröger, R. Visual training improves underwater vision in children. *Vision Research* 46, 3443–3450 (2006).

35. Ilardo, M., et al. Physiological and genetic adaptations to diving in sea nomads. *Cell* 173, 569–580.e15 (2018).

36. Park, D., and Huang, C. Culture wires the brain. *Perspectives on Psychological Science* 5, 391–400 (2010).

37. Blais, C., Jack, R., Scheepers, C., Fiset, D., and Caldara, R. Culture shapes how we look at faces. *PLoS ONE* 3, e3022 (2008).

38. Nisbett, R., Peng, K., Choi, I., and Norenzayan, A. Culture and systems of thought: Holistic versus analytic cognition. *Psychological Review* 108, 291–310 (2001).

39. However, in some places they persist. Remote, rugged Hokkaido island in Japan was settled by former Samurai under the guidance of American agriculturalists, and the frontier spirit still shapes the norms of its citizens 150 years later – they are more individualistic, and less collectivist; more American in behaviour than Japanese.

40. Oota, H., Settheetham-Ishida, W., Tiwawech, D., Ishida, T., and Stoneking, M. Human mtDNA and Y-chromosome variation is correlated with matrilocal versus patrilocal residence. *Nature Genetics* 29, 20–21 (2001).

41. Cohen, D., Nisbett, R., Bowdle, B., and Schwarz, N. Insult, aggression, and the southern culture of honor: An 'experimental ethnography.' *Journal of Personality and Social Psychology* 70, 945–960 (1996).

42. Ellett, W. The death of dueling. *Historia* 59–67 (2004).

43. The Quran doesn't ascribe punishment for gay transgressions and, anyway, there must be four independent witnesses to an act of anal sex. Elsewhere, beautiful young male 'cup bearers' are described as awaiting the faithful in heaven.

44. Saudi religious police target 'gay rainbows.' *The France 24 Observers* (2015). https://observers.france24.com/en/20150724-saudi-police-rainbows-gay-school.

10. TRINKETS AND TREASURES

1. Modern-day Haiti and Dominican Republic.

2. This enterprise cost the lives of 8 million indigenous and African people.

3. Modern-day Indonesia.

4. Ricardo, D. *On the principles of political economy and taxation* (John Murray, 1817).

5. When group size and structure, and benefits and costs, all align, cooperation leads to division of labour, as biological systems such as genes or cells demonstrate.

6. Burk, C. The collecting instinct. *Pedagogical Seminary* 7, 179–207 (1900).

7. Gelman, S., Manczak, E., and Noles, N. The non-obvious basis of ownership: preschool children trace the history and value of owned objects. *Child Development* 83, 1732–1747 (2012).

8. Hood, B., and Bloom, P. Children prefer certain individuals over perfect duplicates. *Cognition* 106, 455–462 (2008).

9. Vanhaereny, M. Middle Paleolithic shell beads in Israel and Algeria. *Science* 312, 1785–1788 (2006).

10. There are many biological systems and ecologies whose evolution could be considered as operating at a group level, such as the gut microbiome.

11. Findeiss, F., and Hein, W. LionMan 2.0 – the experiment. YouTube (2014). https://youtu.be/hgbvT9_pjzo.

12. The harsh environmental conditions of the last ice age pushed our physically weak, furless human ancestors to the limits of survival on a

few occasions – at one point, estimates put our entire population as down to fewer than 10,000 people. That would make us an endangered species, numbering fewer than the chimps around today.

13. Sapiens increased the carrying capacity of the environment by a factor of ten over Neanderthals.

14. The oldest musical instrument we've found, the Hohle Fels flute, made from the wing bone of a griffon vulture, dates to 35,000 years ago.

15. Clarkson, C., et al. Human occupation of northern Australia by 65,000 years ago. *Nature* 547, 306–310 (2017).

16. Liu, W., et al. The earliest unequivocally modern humans in southern China. *Nature* 526, 696–699 (2015). Although this may have been an earlier dispersal that went extinct.

17. They may have survived for so long here by burning the fatty bones of large animals.

18. Beall, C. Two routes to functional adaptation: Tibetan and Andean high-altitude natives. *Proceedings of the National Academy of Sciences* 104, 8655–8660 (2007).

19. Across Africa, skin tone varies greatly and has done so for at least 900,000 years. Many of the genes involved in pigmentation shades evolved in Africa well before modern humans left the continent: see Crawford, N., et al. Loci associated with skin pigmentation identified in African populations. *Science* 358, eaan8433 (2017). San hunter-gatherer people of Botswana, for example, owe their paler skins to the same variant of a gene that noticeably lightens the skin of many Europeans. European light skin owes its provenance to the distribution of these African-evolved genes, as well as some newly acquired ones. Neanderthal populations came in a variety of skin shades, and some of their genes associated with both lighter and darker pigmentation have made it into modern European genomes: Dannemann, M., and Kelso, J. The contribution of Neanderthals to phenotypic variation in modern humans. *American Journal of Human Genetics* 101, 578–589 (2017).

20. The most common pale-skin variant found in Europe is a mutation that arrived only 29,000 years ago in the Near East. This variant entered East Africa, including Tanzania and Ethiopia, with migratory farmers, and also spread north through Europe up to Scandinavia and Scotland, becoming common only a few thousand years ago.

21. It was thought that Africans arrived in Europe, where they needed to wear clothes, and as a consequence of having so little skin exposed and the weaker UV rays, genes for lighter skin were selectively adopted. The beauty of the new tools to analyse ancient DNA is that we can

actually test these theories. And it turns out that things were a little different.

22. Olalde, I., et al. Derived immune and ancestral pigmentation alleles in a 7,000-year-old Mesolithic European. *Nature* 507, 225–228 (2014).

23. The original Indo-European vocabulary, as reconstructed by linguists, contains five words relating to the wheel, revealing it was an important technology for the Yamnaya. Two of these words literally mean 'wheel', one means 'axle', one refers to a pole used to harness animals to a cart, and one is a verb for the action of transporting in a vehicle. These words make it possible to date proto-Indo-European to about 5,500 years ago. That is when full-size wheels, miniature models, and images and carvings of wheeled wagons start to appear across western Eurasia, from the Russian steppes to Poland and all the way down into Mesopotamia.

24. Long, T., Wagner, M., Demske, D., Leipe, C., and Tarasov, P. Cannabis in Eurasia: Origin of human use and Bronze Age trans-continental connections. *Vegetation History and Archaeobotany* 26, 245–258 (2016).

25. In 1997, a hemp rope dating back to 26,900 BCE was found in Czechoslovakia – the first known use of marijuana.

26. The gene codes for the lactase enzyme that breaks down the lactose in milk.

27. Drinking milk as a lactose-intolerant adult results in diarrhoea and stomach cramps.

28. In Chile, populations are today evolving the ability to drink goat milk.

29. In tropical regions, the protective advantage of melanin outweighs any other selective advantage, but different skin tones have emerged across the world in different populations. Recent analysis shows that dark-skinned people have genes that help vitamin D transportation around the body, and northern Europeans have genes that increase vitamin absorption in the skin. The gene for lactase-persistence has also emerged in a handful of other populations, globally.

30. Kristiansen, K., et al. Re-theorising mobility and the formation of culture and language among the Corded Ware culture in Europe. *Antiquity* 91, 334–347 (2017).

31. Rascovan, N., et al. Emergence and spread of basal lineages of Yersinia pestis during the Neolithic decline. *Cell* 176, 295–305.e10 (2019).

32. Goldberg, A., Günther, T., Rosenberg, N., and Jakobsson, M. Ancient X chromosomes reveal contrasting sex bias in Neolithic and Bronze Age Eurasian migrations. *Proceedings of the National Academy of Sciences* 114, 2657–2662 (2017).

33. Yamnaya are associated with a cult of wolves and dogs; they are often buried in wolfskin with canine teeth around their necks.

34. Romulus and Remus, the protagonists in the founding myth of Rome, may be based on Yamnaya insurgents wearing wolf skins and teeth necklaces.

35. www.nature.com/articles/nature25738.

36. Such as the Botain.

37. The earliest evidence of silk – a cocoon cut with a knife and fragments of a loom – dates to around 6,000 years ago.

38. Curiously, shell beads disappear from the archaeological record in Africa and the Near East 70,000 years ago, along with other cultural innovations such as engravings on ochre slabs, and refined bone tools and projectile points. They reappear in different forms up to 30,000 years later, with personal ornaments simultaneously re-emerging in Africa and the Near East, and for the first time in Europe and Asia. This may reflect an entirely new and independent phase of population growth with innovations that allowed a more efficient exploitation of a wider variety of environments. The temporary disappearance of cultural innovations could be linked to population decreases during a period of harsher climate conditions 60,000 to 73,000 years ago. This would have isolated populations, disrupting social and exchange networks.

39. Collard, M., Buchanan, B., and O'Brien, M. Population size as an explanation for patterns in the Paleolithic archaeological record. *Current Anthropology* 54, S388–S396 (2013).

40. Raghavan, M., et al. The genetic prehistory of the New World Arctic. *Science* 345, 1255832–1255832 (2014).

41. Kline, M., and Boyd, R. Population size predicts technological complexity in Oceania. *Proceedings of the Royal Society B: Biological Sciences* 277, 2559–2564 (2010).

42. From my own global travels, I have seen the difference that a road or an Internet connection can make by instantly connecting one group to a great many, and the rapid acceleration in new technologies and opportunities that result. And I have seen its reverse, the isolation resulting from conflict or natural disaster, that reduces a population in size and changes its demography, robbing it of young (predominantly) men – villages that once had machinery and enterprise reduced to clusters of desperate people digging the earth with the most basic hand tools.

43. Henrich, J. *The secret of our success* (Princeton University Press, 2015).

44. Muthukrishna, M., Shulman, B., Vasilescu, V., and Henrich, J. Sociality influences cultural complexity. *Proceedings of the Royal Society B: Biological Sciences* 281, 20132511–20132511 (2013).

45. Derex, M., Beugin, M., Godelle, B., and Raymond, M. Experimental evidence for the influence of group size on cultural complexity. *Nature* 503, 389–391 (2013).

46. A study by anthropologists to settle the academic dispute was unable to decide either way, and debate continues to rage. Derex, M., Beugin, M., Godelle, B., and Raymond, M. Derex et al. Reply. *Nature* 511, E2–E2 (2014).

47. Dalgaard, C., Kaarsen, N., Olsson, O., and Selaya, P. Roman roads and persistence in development | VOX, CEPR Policy Portal. Voxeu.org (2018). https://voxeu.org/article/roman-roads-and-persistence-development.

48. Even into the twentieth century, taxes could be paid in cowry shells in Uganda.

49. On 8 November 2016, India's prime minister Narendra Modi gave just four hours' notice that virtually all the cash in the world's seventh-largest economy would be effectively worthless. The government's 'demonetization' experiment was an attempt to crack down on tax evasion by people stockpiling cash (only around 1 per cent of the population pays income tax). The initiative was a disaster. With 1,000- and 500-rupee notes declared invalid, some 86 per cent of the country's circulating cash was banned overnight in an economy that runs 90 per cent on cash, including house sales and food. Long queues formed at banks, the police were called to manage the crowds, businesses that no longer had cash to pay workers were forced to close and the developing economy slowed and contracted. It was too soon and sudden for India, and was cancelled after a few months, but in time the country will follow the digital path of other economies.

II. BUILDERS

1. A recently uncovered burial of a prestigious man, who lived some 34,000 years ago, was discovered at Sungir in Russia – his body had been decorated with 25 mammoth ivory bracelets and 3,000 ivory beads.

2. Similar monuments have been found globally, including at Turkana in Kenya. Hildebrand, E., et al. A monumental cemetery built by eastern Africa's first herders near Lake Turkana, Kenya. *Proceedings of the National Academy of Sciences* 115, 8942–8947 (2018).

3. The first alcohol may well have been made from fermented fruit rather than grains, and date back hundreds of thousands of years to the first

wooden vessels. I will never forget coming across a fig tree in a tropical forest, bearing overripe and fallen figs, and surrounded by a party of drunk animals, from monkeys to wild pigs.

4. It was perhaps in Europe that humanity's environmental transformations extended to adding (rather than extinguishing) a new species: creating a domestic dog from the wild European grey wolf, gaining protection, company and even warmth on icy nights. It could have been achieved within a human lifetime: a famous experiment, started in 1959 by the Russian scientist Dmitry Belyaev, demonstrated that selectively breeding the tamest wild foxes produced a new tame variety within decades. By 30 generations of breeding, half of the population was tame; by 2006, virtually all were. These 'artificial' domesticates were physically different too, with floppy ears, different coat colourings, wagging tales and the ability to seek and hold human eye contact. It turned out that the genes for aggressiveness is inherited along with the genes for other traits, like coat colour. Hunting with dogs may have helped give our European ancestors the edge over Neanderthals.

5. Bocquet-Appel, J. When the world's population took off: The springboard of the Neolithic demographic transition. *Science* 333, 560–561 (2011).

6. One such village, Ain Ghazal, occupied 10,000 years ago in present-day Jordan, had stone homes with white plastered walls and timber roof beams. Its inhabitants worshipped at circular shrines with metre-tall, wide-eyed sculptures, and buried their dead under their houses, decapitating them to decorate the skulls.

7. Coulson, S., Staurset, S., and Walker, N. Ritualized behavior in the Middle Stone Age: Evidence from Rhino Cave, Tsodilo Hills, Botswana. *PaleoAnthropology* 18–61 (2011).

8. Sage, R. Was low atmospheric CO_2 during the Pleistocene a limiting factor for the origin of agriculture? *Global Change Biology* 1, 93–106 (1995).

9. Our ancestors' digestive systems diversified according to the native and artificial plant types they ate, adapting to improve survival rates. Current populations show distinct genetic differences according to whether or not their ancestors were grain farmers. The cultural evolution to alcohol production also produced genetic changes in some populations. In China, where people have been fermenting a rice wine drink for some 9,000 years, a large proportion of the population (99 per cent in southeastern China) have a gene variant that enables more efficient metabolism of alcohol. This gene (ADH1B) reduces alcohol addiction, but it comes with some pretty miserable side effects, including worse hangovers and flushing, nausea and dizziness when boozing (because it changes

the way the liver breaks down alcohol, increasing the amount of acetaldehyde).

10. Shennan, S., et al. Regional population collapse followed initial agriculture booms in mid-Holocene Europe. *Nature Communications* 4 (2013).

11. Kohler, T., et al. Greater post-Neolithic wealth disparities in Eurasia than in North America and Mesoamerica. *Nature* 551, 619–622 (2017).

12. Çatalhöyük research project. *Çatalhöyük Research Project* (2019). http://www.catalhoyuk.com.

13. Boserup, E. *Woman's role in economic development* (George Allen and Unwin Ltd., 1970).

14. Holden, C., and Mace, R. Spread of cattle led to the loss of matrilineal descent in Africa: A coevolutionary analysis. *Proceedings of the Royal Society of London. Series B: Biological Sciences* 270, 2425–2433 (2003).

15. Alesina, A., Giuliano, P., and Nunn, N. On the origins of gender roles: Women and the plough. *Quarterly Journal of Economics* 128, 469–530 (2013).

16. Talhelm, T., et al. Large-scale psychological differences within China explained by rice versus wheat agriculture. *Science* 344, 603–608 (2014).

17. Talheim Death Pit in Germany contains the 7,000-year-old remains of 34 early farmers – men, women, and children – all apparently axed in the skull and hastily buried.

18. In small-scale societies, as many as 60 per cent of male deaths are attributable to warfare.

19. Although the Incas managed to develop a potato-based tax system.

20. In an attempt to fill an economic hole, the Emperor Vespasian even slapped a urine tax on tanneries.

21. A comparable example of this occurred in tenth-century Iceland, when a devastating volcanic eruption was interpreted by the Viking settlers as a 'sign' and used to herald their conversion to Christianity.

22. It involved a competition among the islanders to bring back the first egg of a particular seagull species.

23. However, the islanders were more than helped to their fate by Europeans who discovered the island in the nineteenth century. Slave raiders from Peru took most of the adults in the 1860s, including the only people capable of writing the Rapa Nui Rongorongo script, the only example of a Polynesian written language. With no one left who could read or write, the script became instantly and forever indecipherable. To add to the tragedy, when the slave raiders were forced to repatriate people to the island, they did so knowingly including people infected with smallpox, thereby reducing the island's population to levels where the dead

could not be buried. By 1870, 97 per cent of the islands population was dead, leaving just 111 individuals.

24. Kohler, T., et al. Greater post-Neolithic wealth disparities in Eurasia than in North America and Mesoamerica. *Nature* 551, 619–622 (2017).

25. Basu, A., Sarkar-Roy, N., and Majumder, P. Genomic reconstruction of the history of extant populations of India reveals five distinct ancestral components and a complex structure. *Proceedings of the National Academy of Sciences* 113, 1594–1599 (2016).

26. In the churches, they had to use their own doors (at least 60 Pyrenean churches still have 'Cagot' entrances) and their own fonts, and they were given communion on the end of long wooden spoons. When a Cagot came into a town, they had to report their presence by shaking a rattle, like a leper ringing his bell. Cagots were forbidden to enter most trades or professions – many ended up as coffin makers. They were not allowed to walk barefoot, like normal peasants, which gave rise to a legend that they had webbed toes. When they went about, they had to wear a goose's foot conspicuously pinned to their clothes. The Cagots weren't even allowed to eat alongside non-Cagots, nor share their dishes, and punishments included having a hand chopped off and nailed to the church door.

27. Kanngiesser, P., and Warneken, F. Young children consider merit when sharing resources with others. *PLoS ONE* 7, e43979 (2012).

28. Washinawatok, K., et al. Children's play with a forest diorama as a window into ecological cognition. *Journal of Cognition and Development* 18, 617–632 (2017).

29. Donnell, A., and Rinkoff, R. The influence of culture on children's relationships with nature. *Children, Youth and Environments* 25, 62 (2015).

30. In an unusual acknowledgement of this, new laws in New Zealand agreed with Maori communities, giving a national park and river system legal personhood.

31. Broushaki, F., et al. Early Neolithic genomes from the eastern Fertile Crescent. *Science* 353, 499–503 (2016).

32. Leslie, S., et al. The fine-scale genetic structure of the British population. *Nature* 519, 309–314 (2015).

33. This is because of genetic drift: they left few descendants or they've not been sampled yet. This is why multiple strands of evidence are useful from archaeology to genetics in understanding ancient movements.

34. Novembre, J., et al. Genes mirror geography within Europe. *Nature* 456, 98–101 (2008).

35. Prado-Martinez, J., et al. Great ape genetic diversity and population history. *Nature* 499, 471–475 (2013).

36. Rohde, D., Olson, S., and Chang, J. Modelling the recent common ancestry of all living humans. *Nature* 431, 562–566 (2004).

37. Probably via the daughter of the Emir of Seville.

38. Hellenthal, G., et al. A Genetic atlas of human admixture history. *Science* 343, 747–751 (2014). And have a play on this: World ancestry. Admixturemap.paintmychromosomes.com (2014). http://admix turemap.paintmychromosomes.com.

39. Duncan, S., Scott, S., and Duncan, C. J. J. Reappraisal of the historical selective pressures for the CCR5-32 mutation. *Journal of Medical Genetics* 42, 205–208 (2005).

40. Jones, S. Steve Jones on Extinction. Edge.org (2014). https://www.edge.org/conversation/steve_jones-steve-jones-on-extinction.

41. Robb, G. *The discovery of France* (Macmillan, 2007).

42. On a visit to Paris in 1254, King Henry III of England was much struck with 'the elegance of the houses which were made of plaster, and were three-chambered and even of four or more storeys (stationum).' Cited in Salzman, L. *Building in England* (Clarendon Press, 1952).

43. Bettencourt, L., Lobo, J., Helbing, D., Kuhnert, C., and West, G. Growth, innovation, scaling, and the pace of life in cities. *Proceedings of the National Academy of Sciences* 104, 7301–7306 (2007).

44. van Dorp, L., et al. Genetic legacy of state centralization in the Kuba Kingdom of the Democratic Republic of the Congo. *Proceedings of the National Academy of Sciences* 116, 593–598 (2018).

45. Lead is easily absorbed into the body and causes a range of health problems, including anaemia, IQ reduction and behavioural problems. Exposure to lead during pregnancy reduces the baby's head size. In the twentieth century, lead was routinely added to petrol and was associated with a significant rise in crime and antisocial behaviour during that time.

46. Harper, K. *The fate of Rome: Climate, disease, and the end of an empire* (Princeton University Press, 2017).

47. It's interesting to note that Christianity is the only major religion without teachings on hygiene.

48. When hot weather exacerbated the stench of raw sewage filling the River Thames.

49. Vassos, E., Pedersen, C., Murray, R., Collier, D., and Lewis, C. Meta-analysis of the association of urbanicity with schizophrenia. *Schizophrenia Bulletin* 38, 1118–1123 (2012).

50. Peen, J., Schoevers, R., Beekman, A., and Dekker, J. The current status of urban-rural differences in psychiatric disorders. *Acta Psychiatrica Scandinavica* 121, 84–93 (2010).

51. Kubota, T. Epigenetic alterations induced by environmental stress associated with metabolic and neurodevelopmental disorders. *Environmental Epigenetics* 2, dvw017 (2016).

52. Serpeloni, F., et al. Grandmaternal stress during pregnancy and DNA methylation of the third generation: An epigenome-wide association study. *Translational Psychiatry* 7, e1202 (2017).

12. TIMEKEEPERS

1. Foer, J. Caveman: An interview with Michel Siffre. *Cabinet Magazine* (2008). http://www.cabinetmagazine.org/issues/30/foer.php.

2. Many of our illnesses are rhythmic too, including the activities of pathogens and cancers.

3. Sleep timetables vary with age, with teens going to bed late and rising late, while the elderly do the opposite – perhaps a survival adaptation to group living, ensuring that at least one person in a mixed-age camp would be on guard during the night.

4. Research on scrub jays and squirrel monkeys show they may have some elements of foresight.

5. Young, J., et al. A theta band network involving prefrontal cortex unique to human episodic memory (2017). doi:10.1101/140251.

6. Templer, V., and Hampton, R. Episodic memory in nonhuman animals. *Current Biology* 23, R801–R806 (2013).

7. For more on this, C. Hammond's *Time warped: Unlocking the mysteries of time perception* (Canongate, 2012) is excellent.

8. Painted on the wall on the Shaft of the Dead Man cave is a bull, a birdman and a bird on a stick – their outlines, with the eyes of the bull, represent the three bright stars of the northern hemisphere summer months: the Summer Triangle, of Vega, Deneb and Altair. At the time they were painted, this region of sky would never have set below the horizon and would have been especially prominent at the start of spring. Nearer to the entrance of this cave complex is a magnificent painting of a bull with a map of the Pleiades cluster of stars hanging over its shoulder. Inside the bull painting, there are spots that may represent other stars found in the region that, today, form part of the constellation of Taurus the bull.

9. Lynch, B., and Robbins, L. Namoratunga: The first archeoastronomical evidence in sub-Saharan Africa. *Science* 200, 766–768 (1978).

10. The Wurdi Youang ring of rocks, dating back more than 11,000 years, maps equinoxes and solstices, and the changing positions of the setting

sun throughout the year. Evidence of terraces near the site even suggests that this observatory may once have supported farming communities that needed accurate seasonal calendars.

11. Arnhem Land Aboriginal stories explain the relationship between the moon and tides: when the tides are high, water fills the moon as it rises; as the water runs out of the moon, the tides fall, leaving the moon empty for three days; then the tide rises once more, refilling the moon.

12. I am writing this at 10 PM in June 2018, but according to my Ethiopian friend Mesi, it is currently the tenth month (of 13) of the year 2010. In the Hebrew calendar it is the year 5778 and it's 4 AM because the Hebrew day starts and ends at sunset rather than midnight; it's the tenth month (Shawwal) of the year 1439 in the Islamic Hijri calendar; for the Hindus, it is 4 AM in the year 5119; for the Chinese it is the year 4716 (also the Year of the Dog), and so on.

13. One solution, arrived at variously by the ancient Chinese, Babylonians, and medieval Swedes, was a 19-year cycle of seven 13-month years alternating with twelve 12-month years. Because 235 lunar months equate almost exactly to 19 solar years, this calendar would only deviate from the true solar year by one day every 219 years. The Jewish calendar was closely modelled on this, whereas the ancient Egyptians followed a 30-day 12-month year, with a 5-day religious celebration at the end of the year. Ptolemy III suggested the leap-year solution of adding an extra day every four years, which was eventually adopted, 200 years later, to coincide with the Julian calendar that the Romans constructed.

14. Elsewhere, the New Year is marked by celestial events, such as the rising of Sirius for the Egyptians, or natural phenomena such as when the pall worm begins to spawn in the Trobriand Islands of the western Pacific. The Mayans, who kept detailed and complicated calendars and believed they could interact with time and influence its passage, had a year of 365 days, and so the months floated around, even though their calculation of the solar year was more accurate than the one our Gregorian calendar relies on.

15. The introduction in 1582 of Pope Gregory VIII's calendar, intended to reform the Julian calendar, which had slipped away from the true solar year by more than a week, was resisted by non-Catholic Europeans. The new calendar meant losing ten days to catch up, and for sociopolitical reasons, England held out adopting it for well over a century, by which time the nation was out of sync by 11 days.

16. The Romans had an eight-day week, only changing to the Jewish seven-day cycle after Constantine's conversion to Christianity. The Inca of Peru

also had an eight-day week. Indigenous people in Bali, Indonesia, and also in Bogota, Colombia, observed a three-day week; a four-day week was common among West African groups; a five-day week was used by tribes from the Hittites to Mongolians. Ancient China had a ten-day week.

17. Inherited from the Sumerians.

18. The problem with a water clock, apart from needing to keep it topped up with water (the Greek name for it was 'water thief') was that the water pressure needed to be constant to keep the flow constant.

19. 'Wasting time', in Latin, was 'losing water' (aquam pedere).

20. Eostre is the Norse goddess of spring.

21. However, longitude remained impossible to determine aboard ships that ventured far from land, because for that you needed a clock so accurate that port-time could be compared to the new time at your latest position at sea. This longitude problem meant large ocean crossings were risky for the Europeans expanding their knowledge of the continents beyond Eurasia. Around the world, awards were offered to whoever could solve the problem.

22. Although these early, unreliable clocks needed to be reset against true sun-time.

23. The observation, in 1583 by a young Galileo, of a swinging church chandelier in Pisa, and his realization that the time it takes a pendulum to complete its arc depends only on the length of the string, led to the development of a pendulum clock, the most accurate timekeeping device to date. But it took a clock maker from Yorkshire, John Harrison, to design the ultimate pendulum clock in the 1750s. Accurate to the second over 100 days, Harrison's Clock B (H4) revolutionized seafaring. Longitude could be determined far out at sea, new lands were discovered and important scientific breakthroughs resulted: a relationship between location coordinates and time that continues to this day with GPS satellites maintaining clock accuracy.

24. The invention and popularity of the pocket watch, allowing us – perhaps prompting us – to check the time at any second, only hurried us more. Lewis Carroll, who loved to play with our idea of time, had Alice's adventures take place in the time-expanding zone of a dream; she attends a tea party in which the Hatter reveals that they have tea all day because Time has punished him by eternally standing still at 6 PM (tea time). Carroll based his white rabbit on the dean at Christchurch College in Oxford, who used to appear at a little door and take his watch out of his pocket and say, 'I'm late, I'm late, I'm so sorry,' because everyone would be waiting for his arrival to eat their dinner.

25. Levine, R., and Norenzayan, A. The pace of life in 31 countries. *Journal of Cross-Cultural Psychology* 30, 178–205 (1999).

26. But each time a leap second is added, our modern, time-dependent world is upset – in 2012, for instance, the booking systems of several airlines were disrupted by the manipulation of a leap second.

27. 'I think it inevitably follows, that as new species in the course of time are formed through natural selection, others will become rarer and rarer, and finally extinct. The forms which stand in closest competition with those undergoing modification and improvement will naturally suffer most,' he wrote.

28. Even though your own DNA is unique to you, it can be compared to the DNA of a fern, a mammoth, or any of the more than 100 billion humans who have ever lived. This comparative genomics, population genetics, is the clock that allows us to time travel to visit our common ancestors.

29. Our latest scientific understanding of time feels so divorced from our intuitive time that it allows the impossibles of our imagination to also be possible. Models of multiverses, of entanglement or of time flowing backwards don't make intuitive sense, or help us know what the universe is like really. For that, many of us continue to turn to other cultural ways of knowing.

13. REASON

1. Pluchino, A., Biondo, A., and Rapisarda, A. Talent versus luck: The role of randomness in success and failure. *Advances in Complex Systems* 21, 1850014 (2018).

2. Goldman, J. Friday fun: Snowboarding crow [video]. *Scientific American Blog Network* (2019). https://blogs.scientificamerican.com/thoughtful-animal/friday-fun-snowboarding-crow-video/.

3. Kark, S., Iwaniuk, A., Schalimtzek, A., and Banker, E. Living in the city: Can anyone become an 'urban exploiter'? *Journal of Biogeography* 34, 638–651 (2007).

4. Reducing the costs of failure would be a strategy to boost innovation.

5. Miu, E., Gulley, N., Laland, K., and Rendell, L. Innovation and cumulative culture through tweaks and leaps in online programming contests. *Nature Communications* 9 (2018).

6. A ratchet is a tool with angled teeth that allows cumulative, step-wise progress in one direction only, with no regression between steps.

7. The term 'scientist' was not actually used until 1834.

8. Thales's students included Pythagoras and Anaximander, who attempted to draw a world map by compiling reports from people who had travelled outside of the Mediterranean, and explained lightning and thunder as violent collisions of air and clouds, and rain as the fall of evaporated ocean water. Anaximander also postulated an evolutionary explanation for life, speculating that our human ancestor was a fish-like creature. He also theorized that all matter was made of the same stuff, something that was built on a century later by Democritus, a widely travelled mathematician and physicist who proposed an atomic theory of matter, some 2,300 years before Rutherford.

9. Freeman, C. *The closing of the western mind: The rise of faith and the fall of reason* (Alfred A. Knopf, 2003).

10. 'Almost alone, virtually the last academic, she stood for intellectual values, for rigorous mathematics, ascetic Neoplatonism, the crucial role of the mind, and the voice of temperance and moderation in civic life,' wrote the historian Michael Deakin in *Hypatia of Alexandria* (Prometheus Books, 2007).

11. Cinnirella, F., and Streb, J. Religious tolerance as engine of innovation. CESifo Working Paper Series No. 6797 (2018).

12. Baghdad may have been the first city of over a million inhabitants. It was situated on the banks of the Tigris where it flows closest to the Euphrates, and was a natural crossroads for travellers and traders from Europe, Asia and Africa. The scholars of Baghdad were notable for their scientific approach. A remarkable scientist of the age, Ibn al-Haytham, who published a *Book of Optics,* wrote: 'The duty of the man who investigates the writings of scientists, if learning the truth is his goal, is to make himself an enemy of all that he reads, and . . . attack it from every side. He should also suspect himself as he performs his critical examination of it, so that he may avoid falling into either prejudice or leniency.' Experiments in engineering and invention were described in the *Book of Ingenious Devices,* a large illustrated work on mechanical devices that included automata, puzzles and magic tricks; it describes perhaps the earliest example of a programmable machine: a robotic flute player.

13. Assembly-line methods of hand-copying manuscripts allowed publishers to turn out editions far larger than any available in Europe for centuries.

14. There were a few flickering lights, such as the court of Charlemagne, which hosted scholars including the English philosopher Alcuin (inventor of the wolf, goat and cabbage river-crossing mathematical problem), and the court of Alfred the Great, who championed education. But it wasn't

until the twelfth century that important advances were made in science and technology.

15. The blinkering of Christianity to reasoning and inquiry, however, is only a partial story. Belief in a spiritual or religious agency needn't preclude a scientific approach, and many of the world's scientists have pursued their curiosity in order to feel closer to their god. Descartes, for example, was driven to determine what of the universe could be known through mathematics, logic and deduction; he viewed this as a religious calling. Although the Catholic Church has been hostile to scientific inquiry, it has also been one of the strongest supporters of precision astronomy. Dozens of churches and cathedrals across Europe were also observatories throughout the Middle Ages, many with strategically placed apertures to enable a beam of sunlight to strike a north–south meridian line on the floor.

16. With new forms of classification, publishers established the lasting division between sciences and humanities so as to divide their catalogues more easily. Instead of commenting on a few canonical texts, intellectuals learned to navigate whole libraries of information. In the process they invented the modern idea of the fact: reliable information that could be checked and tested. Suddenly, for instance, Montaigne 'could see more books [in] a few months ... than earlier scholars had seen after a lifetime of travel', and consequently conflict, diversity and contradictions became more visible to him than to his predecessors.

17. As books became cheaper, more portable, and therefore more desirable, they expanded the market for all publishers, heightening the value of literacy still further. There was a proliferation of print shops issuing authoritative, widely tangible information.

18. At the same time, it spurred a new religious fundamentalism, as more people became conversant with the literal Bible.

19. Like Marsilio Ficino and Pico della Mirandola.

20. The Royal Society was founded in 1660, dedicated to experimentation and concerned with 'facts not explanations'.

21. Heyes, C. Grist and mills: On the cultural origins of cultural learning. *Philosophical Transactions of the Royal Society B: Biological Sciences* 367, 2181–2191 (2012).

22. Henrich, J. Why societies vary in their rates of innovation: The evolution of innovation-enhancing institutions. *Innovation in Cultural Systems: Contributions from Evolutionary Anthropology*, Altenberg Workshops in Theoretical Biology, Konrad Lorenz Institute, Altenberg, Austria (2007). Available at https://pdfs.semanticscholar.org/8684/a4f1b3eae05dcff3ba1f03c5678c3359c215.pdf.

23. Muthukrishna, M., and Henrich, J. Innovation in the collective brain. *Philosophical Transactions of the Royal Society B: Biological Sciences* 371, 20150192 (2016).

24. Mackey, A., Whitaker, K., and Bunge, S. Experience-dependent plasticity in white matter microstructure: Reasoning training alters structural connectivity. *Frontiers in Neuroanatomy* 6 (2012); and Qin, Y., et al. The change of the brain activation patterns as children learn algebra equation solving. *Proceedings of the National Academy of Sciences* 101, 5686–5691 (2004).

25. Stanovich, K. Rational and irrational thought: The thinking that IQ tests miss. *Scientific American Mind* 20, 34–39 (2009); and Bloom, P., and Weisberg, D. Childhood origins of adult resistance to science. *Science* 316, 996–997 (2007).

26. Fears of this last has excluded women and other groups from intellectual debate, and their ideas and inventions have been misattributed. The false rhetoric that they were not capable of logical reasoning was widely believed even by those maligned. This generates a feedback loop, because being socially marginalized has a measurable impact on cognition, learning and achievement.

27. Frank, M., and Barner, D. Representing exact number visually using mental abacus. *Journal of Experimental Psychology: General* 141, 134–149 (2012).

28. Thomas Newcomen, who invented the steam engine, was an ironmonger.

29. One billion is 1,000 millions; 1 trillion is 1 million millions. One million seconds is approximately 12 days; 1 billion seconds is approximately 32 years.

30. We also lose a host of other infant abilities, such as being able to distinguish between different monkey faces.

31. We use these platforms to talk to ourselves or to others who think like us, have the same perception of reality, and don't question ours. It is making us more extreme in our views, less tolerant of other views and more easy to manipulate.

32. Filipowicz, A., Barsade, S., and Melwani, S. Understanding emotional transitions: The interpersonal consequences of changing emotions in negotiations. *Journal of Personality and Social Psychology* 101, 541–556 (2011).

33. Kanai, R., Feilden, T., Firth, C., and Rees, G. Political orientations are correlated with brain structure in young adults. *Current Biology* 21, 677–680 (2011).

34. Block, J., and Block, J. Nursery school personality and political orientation two decades later. *Journal of Research in Personality* 40, 734–749 (2006).

35. Nail, P., McGregor, I., Drinkwater, A., Steele, G., and Thompson, A. Threat causes liberals to think like conservatives. *Journal of Experimental Social Psychology* 45, 901–907 (2009).

36. Huang, J., Sedlovskaya, A., Ackerman, J., and Bargh, J. Immunizing against prejudice. *Psychological Science* 22, 1550–1556 (2011).

37. Napier, J., Huang, J., Vonasch, A., and Bargh, J. Superheroes for change: Physical safety promotes socially (but not economically) progressive attitudes among conservatives. *European Journal of Social Psychology* 48, 187–195 (2017).

38. Harrington, J., and Gelfand, M. Tightness-looseness across the 50 United States. *Proceedings of the National Academy of Sciences* 111, 7990–7995 (2014).

39. Gelfand, M., et al. Differences between tight and loose cultures: A 33-nation study. *Science* 332, 1100–1104 (2011).

40. This is how identity politics can become the enemy of reason.

41. Newport, F., and Dugan, A. College-educated Republicans most skeptical of global warming. Gallup (2015). http://news.gallup.com/poll/182159/college-educated-republicans-skeptical-global-warming.aspx.

42. Oak Ridge National Laboratory launches America's new top supercomputer for science. US Department of Energy (2018). https://www.energy.gov/articles/oak-ridge-national-laboratory-launches-america-s-new-top-supercomputer-science.

14. HOMNI

1. Introduction: 10,000 Year Clock – the long now. Longnow.org (2019). http://longnow.org/clock/.

2. This may be behind the finding that wild primates are far less curious and innovative than captive ones.

3. Runco, M., Acar, S., and Cayirdag, N. Further evidence that creativity and innovation are inhibited by conservative thinking: Analyses of the 2016 presidential election. *Creativity Research Journal* 29, 331–336 (2017).

4. Mani, A., Mullainathan, S., Shafir, E., and Zhao, J. Poverty impedes cognitive function. *Science* 341, 976–980 (2013).

5. Ziegler, M., et al. Development of Middle Stone Age innovation linked to rapid climate change. *Nature Communications* 4 (2013).

6. The Roman Empire used water wheels, oil and coal, but largely relied on human slave labour – as much as 40 per cent of the population were

slaves. A shortfall in the supply of slaves in around 150 CE contributed to the collapse of the civilization to sustainable levels.

7. Nordhaus, W. Do real output and real wage measures capture reality? The history of lighting suggests not. *Economics of New Goods* 58, 29–66 (1997).

8. Fouquet, R., and Pearson, P. The long run demand for lighting: Elasticities and rebound effects in different phases of economic development. *Economics of Energy & Environmental Policy* 1 (2012). It's nicely visualized here: The price for lighting (per million lumen-hours) in the UK in British pounds. *Our World in Data* (2012). https://ourworldindata.org/grapher/the-price-for-lighting-per-million-lumen-hours-in-the-uk-in-british-pound.

9. Compared to 3 per cent in 1800.

10. Hellenthal, G., et al. A genetic atlas of human admixture history. *Science* 343, 747–751 (2014).

11. A century ago, average life expectancy was around 50; now it's more than 80. The average number of children a woman had in 1800 was six, now it's approaching two (and less than that in some countries). Meanwhile, culturally produced social inequalities are having real biological effects. Globally, one in five children is stunted, including nearly 40 per cent of Indian children.

12. But this rapid pace of innovation means the half-life for knowledge – the time it takes for expertise to become obsolete – is ever shrinking.

13. Steffen, W., et al. *Global change and the earth system: A planet under pressure* (Springer, 2004).

14. Although China, Russia and others are increasingly erecting digital borders, attempting to assert Internet sovereignty within their geopolitical borders.

15. Global citizenship a growing sentiment among citizens of emerging economies shows global poll for BBC World Service – Media Centre. BBC (2016). https://www.bbc.co.uk/mediacentre/latestnews/2016/world-service-globescan-poll.

16. Rozin, P. The weirdest people in the world are a harbinger of the future of the world. *Behavioral and Brain Sciences* 33, 108–109 (2010).

17. A baby born today in the industrialized world, in Homni's Anthropocene, could be described as being born with a kind of environmental Original Sin, because she will add to the unsustainable exploitation of the natural world during her lifetime, worsening conditions for everyone. Or she could prove our salvation. Our hypercooperation means we can collectively close a hole in the stratospheric ozone layer, but also that we can make it in the first place.

18. This is certainly the case in the unequal cities of the United States; it is also the case for the relatively equal Scandinavian cities.

19. Krausmann, F., et al. Global human appropriation of net primary production doubled in the 20th century. *Proceedings of the National Academy of Sciences* 110, 10324–10329 (2013).

20. As we enter a period of global warming, with increasingly limited freshwater and mineral resources, our culture will need to transform from one that consumes water, fuels and materials to one that circulates resources within Homni's global factory, ending the linear production-to-waste model we've used for the past millennia.

21. The Inuit have a word for the environmental changes they are seeing: *uggianaqtuq*, meaning 'to behave strangely'.

22. Levine, H., et al. Temporal trends in sperm count: A systematic review and meta-regression analysis. *Human Reproduction Update* 23, 646–659 (2017).

23. Ralston, J., et al. Time for a new obesity narrative. *The Lancet* 392, 1384–1386 (2018).

24. However, in many parts of the world, people are still not getting adequate protein and micronutrients, so their brains cannot reach full potential, which has biological, social and cultural consequences.

25. Foster, P., and Jiang, Y. Epidemiology of myopia. *Eye* 28, 202–208 (2014). Every additional year in education increases myopia: Mountjoy, E., et al. Education and myopia: Assessing the direction of causality by Mendelian randomisation. *BMJ* k2022 (2018).

26. 40,000,000,000,000,000,000,000,000,000.

27. 'Forgetfulness in the learners' souls, because they will not use their memories . . . they will be hearers of many things and will have learned nothing; they will appear to be omniscient and will generally know nothing; they will be tiresome company, having the show of wisdom without the reality.' Excerpt of a dialogue between Socrates and Phaedrus from *The Phaedrus,* written down by Plato (a pupil of Socrates), in approximately 370 BCE.

28. Must, O., and Must, A. Speed and the Flynn effect. *Intelligence* 68, 37–47 (2018); and Clark, C., Lawlor-Savage, L., and Goghari, V. The Flynn effect: A quantitative commentary on modernity and human intelligence. *Measurement: Interdisciplinary Research and Perspectives* 14, 39–53 (2016).

29. Although some of the largest famines of the twentieth century were caused by politicians championing anti-scientific dogma.

30. There have been at least 180 civil wars since 1960.

Index